Mungbean

Sivakumar Paramasivam • Jing Tian
Muthiyan Pandiyan • Changyou Liu
Wang Lixia
Editors

Mungbean

Genetics and Genomics Perspectives

Springer

Editors
Sivakumar Paramasivam
Dr M.S. Swaminathan Agricultural College
and Research Institute
Tamil Nadu Agricultural University
Thanjavur, Tamil Nadu, India

Jing Tian
Institute of Cereal and Oil Crops
Hebei Academy of Agricultural
and Forestry Sciences
Shijiazhuang, Hebei, China

Muthiyan Pandiyan
Dr M.S. Swaminathan Agricultural College
and Research Institute
Tamil Nadu Agricultural University
Thanjavur, Tamil Nadu, India

Changyou Liu
Institute of Cereal and Oil Crops
Hebei Academy of Agricultural
and Forestry Sciences
Shijiazhuang, Hebei, China

Wang Lixia
Chinese Academy of Agricultural Sciences
Institute of Crop Sciences
Haidian, Beijing, China

ISBN 978-981-95-7636-4 ISBN 978-981-95-7637-1 (eBook)
https://doi.org/10.1007/978-981-95-7637-1

This Springer imprint is published by the registered company Springer Nature Singapore Pte Ltd.
The registered company address is: 152 Beach Road, #21-01/04 Gateway East, Singapore 189721, Singapore

Preface

Mungbean (*Vigna radiata*), a vital leguminous crop cultivated extensively across Asia, Africa, and other tropical regions, has garnered significant attention due to its nutritional value, and adaptability benefits. Mungbean is an excellent source of easily digestible good quality protein for the growing population in many developing countries of South Asia and Southeast Asia. It is also a rich source of dietary fiber, vitamins, and minerals. Mungbean plays a crucial role in food security and sustainable agriculture.

In recent years, advances in genetics and genomics have opened new avenues for understanding the underlying biological mechanisms that govern Mungbean development, stress tolerance, and yield improvement. The integration of genomic tools such as genome sequencing, molecular marker development, and gene editing has accelerated breeding efforts and facilitated the identification of genes associated with desirable traits.

This compilation, "Mungbean—Genetics and Genomics Perspectives," aims to synthesize current knowledge, innovative research findings, and future directions in the field. It underscores the importance of genomic insights in breeding programs and highlights the potential of genomics-driven approaches to enhance Mungbean productivity, resilience, and nutritional quality. As we delve into this rapidly evolving domain, it becomes evident that harnessing genetic and genomic resources holds the key to unlocking the full potential of Mungbean for a sustainable and food-secure future.

Thanjavur, Tamil Nadu, India — Sivakumar Paramasivam
Shijiazhuang, Hebei, China — Jing Tian
Thanjavur, Tamil Nadu, India — Muthiyan Pandiyan
Shijiazhuang, Hebei, China — Changyou Liu
Haidian, Beijing, China — Wang Lixia

Contents

Editors and Contributors

About the Editors

Sivakumar Paramasivam is presently working as an Professor (Biotechnology) at Dr. M. S. Swaminathan Agricultural College and Research Institute, Tamil Nadu Agricultural University, Eachangkottai, Thanjavur, Tamil Nadu, India. He has completed his graduation at the Faculty of Agriculture, Annamalai University, Chidambaram, Tamil Nadu. He has completed his MSc (Ag.) in Plant Breeding and Genetics and PhD in Biotechnology at Tamil Nada Agricultural University, Coimbatore, Tamil Nadu, India. He did his Post-Doctoral Research Fellow (2008–2010) at the Department of Genetic and Molecular Biology, University of Malaya, Kuala Lumpur, Malaysia. He underwent two months Industrial Development Training at Hebei Academy of Agriculture and Forestry Sciences, China. He has more than 20 years research experience and 15 years teaching experience for teaching both undergraduate and postgraduate students in Agriculture. He has published more than 70 research papers and review articles in reputed national and international journals like *Plant Cell Report*, *Biologia Plantarum*, *Journal of Natural Product Research*, and *Acta Biologica Hungarica*. He has edited 18 books for the benefits of undergraduate and postgraduate students of agriculture and horticultural sciences. He has received several awards for their contribution in grain legumes research. He has also published about 70 popular articles in Tamil and English in monthly science magazines.

Jing Tian is currently working at the Institute of Cereal and Oil Crops, Hebei Academy of Agriculture and Forestry Sciences, Hebei, China. She graduated from Hebei Science and Technology Normal University in 1985, and got the master's degree in Plant Breeding and Genetics from the Chinese Academy of Agricultural Sciences in Beijing, China, in 1999. With 40 years of research experience in food legumes, she bred out 36 new varieties of mungbean and adzuki bean, and has published more than 70 research papers and review articles in renowned national and international journals such as *Plant Communications*, *Theoretical and Applied Genetics*, *Scientia Agricultura Sinica*, *Acta Agronomica Sinica*, etc. As the first contributor, she archived five awards for her contributions to grain legumes research from Hebei government. Now, she is the chief scientist of the Hebei Province Minor Grain and Pulses Breeding Innovation Team.

Muthiyan Pandiyan is presently working as a Professor (PBG) at Dr. M. S. Swaminathan Agricultural College and Research Institute, Tamil Nadu Agricultural University, Eachangkottai, Thanjavur, Tamil Nadu, India. He served in administrative positions for more than eight years as Dean of two Agricultural Colleges and Research Institutes of Tamil Nadu Agricultural University. He also served as Professor and Head of three TNAU research stations and as Programme Coordinator of one ICAR–TNAU Krishi Vigyan Kendra (KVK). He involved, developed and released 17 varieties in different crops namely pulses, oilseeds, vegetable and fruit crops. As Principal Investigator for many research projects worth Rs. 9 crores received to the University funded by National agencies viz., ICAR, DBT, TANII, NABARD and SERB, New Delhi. National pulses permanently scheme of AICRP-Pigeonpea to the TNAU Agriculture Research Station, Virinjipuram, Vellore. Five Indian Registered Patent received. Five geotechnological patents were filed.

He has handled more than 30 courses. He has published 178 research papers in international and national journals, including 11 papers in journals with a NAAS rating of more than 6. In addition, he has published 150 popular articles, authored 12 books, and contributed 20 book chapters. Student Guidance: PG-5, PhD-6. Countries visited and project collaboration with Japan and China, Visited Thailand, Singapore, Indonesia, Malaysia on invitation. Governing member for TNAU four Affiliated Agricultural Colleges. TNAU- Academic Council member and member of board of studies for both UG and PG educations for 2017-2020. PAJANCOA, Karaikal-Pondicherry member for BoS for both Agri and Horti UG educations for 3 years 2022–2025. District level government purchase committee member for three districts. Mobilized 150 crores received from NABARD for all basic infra structures, civil structures created and vehicles purchased in new two Agriculture Colleges and Research Institute as Dean. Organized many state and district level famers mela for Kharif and Rabi seasons.

He has received 44 awards including Norman Borlaug Award 2025 by Agrimeet Foundation, New Delhi; Fellow award 2025 by Tamil Science Society, New Delhi; ICAR-ISPGR award 2023 by NBPGR New Delhi; Life Time Achievement Award by Society of Krishi Vigyan (SKV); Dr.B.P.PAL Memorial NABS Best Scientist award 2021; Dr.Kalayya Krishnamurthy National Award for the Best Agricultural Research 2018–19 by UAS, Bengaluru; Dr.APJ.Abdulkalam Best Scientist Award-2022; TNAU-Best Researcher; TNAU-Best Extension Scientist; ICAR South Zone—Best KVK Vellore, TNAU, and Best–ICAR Research Centre for *Vigna* improvement; and The Best honors for NABS Fellow.

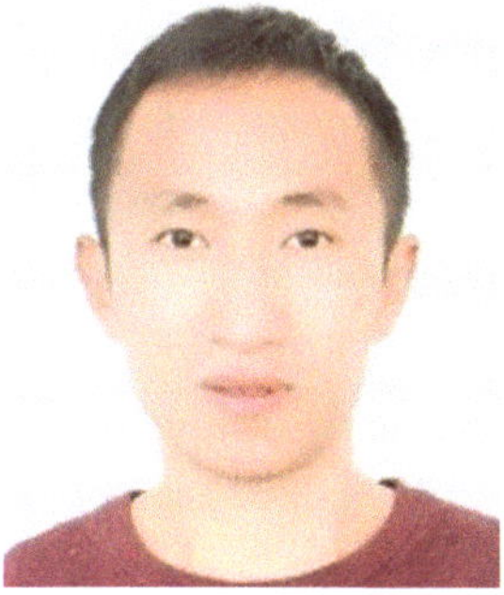

Changyou Liu is currently a Professor at the Institute of Cereal and Oil Crops, Hebei Academy of Agriculture and Forestry Sciences, Hebei, China. He graduated from Qingdao Agricultural University in Qingdao, Shandong province. He earned his MSc (Ag.) in Plant Breeding and Genetics and his PhD in Science of Crop Germplasm Resources from the Chinese Academy of Agricultural Sciences in Beijing, China. With over 20 years of research experience in agriculture, he has published more than 50 research papers and review articles in renowned national and international journals such as *Plant Communications*, *Theoretical and Applied Genetics*, *Scientia Agricultura Sinica*, and *Acta Agronomica Sinica*. He has received several awards for his contributions to grain legumes research.

Wang Lixia is presently working as a Professor (Genetic resources and breeding) at the Institute of Crop Sciences, Chinese Academy of Agricultural Sciences, Beijing, China. She has completed her MSc (Ag.) in the College of Agronomy, Huazhong Agriculture University, and PhD in Biotechnology from the Graduate School of Chinese Academy of Agricultural Sciences, Beijing, China. She was a visiting scholar at the National Institute of Agrobiological Sciences (NIAS) for three months, University of Georgia (UGA) for a whole year, and Kasetsart University (KU, Thailand) for three months. She has more than 20 years research experience on collection and evaluation of genetic resources, and breeding on *Vigna* crops. She has published more than 80 research or review articles in national and international journals like *Nature Communication*, *Crop Journal*, *Planting Breeding*, *Molecular Breeding*, *Crop Science*, *Euphytica*, etc.

Contributors

G. Anand ICAR-KrishiVigyan Kendra, Salem, India

Baojie Fan Institute of Cereal and Oil Crops, Hebei Academy of Agricultural and Forestry Sciences/Hebei Laboratory of Crop Genetic and Breeding, Shijiazhuang, Hebei, China

A. Gopikrishanan Agricultural Research Station, Tamil Nadu Agricultural University, Vellore, Tamil Nadu, India

K. R. Jahanmohan Dr. M.S. Swaminathan Agricultural College and Research Institute, Tamil Nadu Agricultural University, Thanjavur, Tamil Nadu, India

Jothi Kanmani Bharathi Faculty of Agriculture, Annamalai University, Annamalainagar, Tamil Nadu, India

A. Krishanveni Agricultural Research Station, Tamil Nadu Agricultural University, Tiruvannamalai, Tamil Nadu, India

Changyou Liu Institute of Cereal and Oil Crops, Hebei Academy of Agricultural and Forestry Sciences/Hebei Laboratory of Crop Genetic and Breeding, Shijiazhuang, Hebei, China

Wang Lixia Key Laboratory Grain Crop Genetic Resources Evaluation and Utilization, Ministry of Agriculture and Rural Affairs, Institute of Crop Sciences, Chinese Academy of Agricultural Sciences (ICS-CAAS), Beijing, China

A. Mahalingam Regional Research Station, Tamil Nadu Agricultural University, Cuddalore, Tamil Nadu, India

Muthiyan Pandiyan Dr. M.S. Swaminathan Agricultural College and Research Institute, Tamil Nadu Agricultural University, Thanjavur, Tamil Nadu, India

Jianxiang Peng Biobin Data Science, Changsha, China

C. Pradipa Agro Climatic Research Centre, Tamil Nadu Agricultural University, Coimbatore, India

P. Priyadharshini Forest College and Research Institute, Coimbatore, India

V. Radhakrishnan AnbilDharmalingam Agricultural College and Research Institute, Tiruchirappalli, India

Subramanian Rajesh Department of Plant Biotechnology, CPMBB, Tamil Nadu Agricultural University, Coimbatore, Tamil Nadu, India

M. Ravi ICAR-KrishiVigyan Kendra, Salem, India

Kuttralingam Sasikala Dr. M.S. Swaminathan Agricultural College and Research Institute, Tamil Nadu Agricultural University, Thanjavur, Tamil Nadu, India

Sanyasi Selvamani Faulty of Agriculture, Annamalai University, Annamalainagar, Tamil Nadu, India

Yingchao Shen Hebei Key Laboratory of Crop Genetics and Breeding, Shijiazhuang, Hebei, China

Huiying Shi Institute of Cereal and Oil Crops, Hebei Academy of Agricultural and Forestry Sciences, Shijiazhuang, Hebei, China

Sivakumar Paramasivam Dr. M.S. Swaminathan Agricultural College and Research Institute, Tamil Nadu Agricultural University, Thanjavur, Tamil Nadu, India

A. Sowmiya Senior Research Fellow, ICAR-KrishiVigyan Kendra, Sandhiyur, Salem, India

Qiuzhu Su Institute of Cereal and Oil Crops, Hebei Academy of Agricultural and Forestry Sciences/Hebei Laboratory of Crop Genetic and Breeding, Shijiazhuang, Hebei, China

Jing Tian Institute of Cereal and Oil Crops, Hebei Academy of Agricultural and Forestry Sciences, Shijiazhuang, Hebei, China

Shen Wang Institute of Cereal and Oil Crops, Hebei Academy of Agricultural and Forestry Sciences/Hebei Laboratory of Crop Genetic and Breeding, Shijiazhuang, Hebei, China

Xueqing Wang Hebei Academy of Agricultural and Forestry Sciences, Shijiazhuang, Hebei, China

Yan Wang Institute of Cereal and Oil Crops, Hebei Academy of Agricultural and Forestry Sciences/Hebei Laboratory of Crop Genetic and Breeding, Shijiazhuang, Hebei, China

Yanjun Zhang Department of Resource and Environmental Engineering, Xingtai Polytechnic College, Xingtai, Hebei, China

Zhixiao Zhang Institute of Cereal and Oil Crops, Hebei Academy of Agricultural and Forestry Sciences/Hebei Laboratory of Crop Genetic and Breeding, Shijiazhuang, Hebei, China

Genetic Resources Conservation and Assessment

Wang Lixia

Abstract

Genetic resources, also named germplasm resources or germplasm collections, generated with human activities and agricultural origins, are the most important component of biodiversity. Genetic resources are closely related to the survival and development of human being. They are constantly innovating and developing, and are the most fundamental material basis and strategic resources for human reproduction and development. Genetic resources for crop species are also fundamental for breeding and the sustainable development of agriculture. Scientists on crop breeding areas have attached great importance and interests on the collection and preservation of genetic resources, especially during the late of the last century. Thus, there was an upsurge in the collection, conservation, and evaluation of germplasm resources all over the world, and core collections had been constructed for many crops in the past decades, to make a better use and management of collections.

Keywords

Mungbean · Genetic resources · Conservation

W. Lixia (✉)
Key Laboratory Grain Crop Genetic Resources Evaluation and Utilization, Ministry of Agriculture and Rural Affairs, Institute of Crop Sciences, Chinese Academy of Agricultural Sciences (ICS-CAAS), Beijing, China
e-mail: wanglixia03@caas.cn

S. Paramasivam et al. (eds.), *Mungbean*,
https://doi.org/10.1007/978-981-95-7637-1_1

1 Introduction

As a traditional crop in Asia, it is said that mungbean (*Vigna radiata*) is a home crop for Asian countries and has been cultivated in these countries for thousands of years; in recent decades, the cultivation of mungbean has been spread to Africa, Australia, and other continents (Diatta et al. 2024). Germplasm resources for mungbean had been widely collected in many countries too. Till now, China has the largest collection of around 7000 accessions of mungbean, but within this gene pool, germplasm collected from overseas accounts for no more than 10%, and a high genetic redundancy also had been observed using DNA molecular markers. AVRDC (Asian Vegetable Research and Development Center) is an organization that holds the most abundant mungbean resources with about 6700 accessions, followed by National Bureau of Plant Genetic Resources of India (4325 acc.). USDA-USA also conserved about 4000 accessions, and most of were collected from over 40 countries and might hold the highest diversity of all. In addition, Japan, the Philippines, Australia, and Korea conserved about 1000 accession of mungbean germplasm.

1.1 Core Collection Construction and Evaluation

For a better evaluation and utilization of germplasm resources with emphasis, core collections have been constructed for different mungbean germplasm pools. And the first mungbean core collection was established in India based on environment strategy and agronomic traits, and genetic diversity was assessed (Bisht et al. 1998). As the largest mungbean germplasm preservation country, based on the agronomic data from the National genebank, Chinese scientists constructed the primary core collection of mungbean 15 years ago, and this core collection accounted for about 14.2% of total entries and represented 100% of the total phenotypic variations (Liu et al. 2008). Several years later, the core collection of mungbean in China was subsequently constructed, and after evaluation under different environments based on agronomic traits, a series of elite collections suitable for breeding parents or production had been selected (Wang et al. 2014). Additionally, many local core collections were established in China which had been used for GWAS analysis on diverse traits. For instance, based on re-sequencing of 169 landraces and 113 mungbean breeding lines, genomic regions related with drought tolerance were identified (Chang et al. 2023), and based on the resequencing of 558 Chinese landraces, 110 signals that significantly associated with nine agronomic traits were located (Han et al. 2022). While 259 PAV (presence/absence variation) were found to be associated with 33 agronomic traits using 217 accessions (Liu et al. 2022). With the development of high-throughput sequencing technology, whole genome assembly is becoming no longer costive and unacceptable, and we know there are some groups on the world are still working with GWAS analysis on mungbean too (private communications). Based on geographical stratification and cluster analysis of eight phenotypic descriptors, a core and a mini core collection were established in AVRDC mungbean pool, respectively (Schafleitner et al. 2015). In order to get a better use of these two collections, they had been delivered to diverse research groups. Then, agronomic traits (Sokolkova et al. 2020), root phernotypes (Aski et al. 2021), tolerance to waterlogging (Kyu et al. 2024), and salinity (Hanumantha Rao et al. 2016) were continued to be assessed, and GWAS analysis was performed to identify genetic loci or candidate genes useful for breeding. In USA, a core collection representing accessions from 13 countries was evaluated for seed mineral concentration, and a set of loci were identified to be related with seed mineral concentration based on GWAS analysis (Wu et al. 2020). Except for core collections, numerous random sets of mungbean collections had been evaluated for better use of germplasm in production, breeding, or genetic studies. In China, basic traits of mungbean germplasm had been evaluated and cataloged. And lots of studies on local collections had been done based on agronomic traits in the past decades. For instance, Luo et al. (2010) analyzed 112 local mungbean from Guangxi district, and Li et al. (2009) analyzed 338 local mungbean from Hubei province. In addition, small sets of accessions from Shanxi, Heilongjiang, Hebei were also assessed to identify elite germplasm for local breeding and production (Zhang et al. 2013; Gao 2018; Liu et al. 2019; Hou et al. 2020; Lu et al. 2021).

1.2 Evaluation on Resistance and Tolerance

Scientists in AVRDC began to evaluate the resistance of mungbean germplasm from the late of last century, and had identified several important entries that resistant to bruchid, which had played great roles in development of bruchid resistant cultivars. And the erect and early mature lines had been introduced into different countries to enhance the production of local mungbean. Zhu (2018) established a method for identifying resistance to fusarium wilt of mungbean and discovered 46 resistant lines. Lang et al. (2024) evaluated the tolerance of 35 Mungbean lines at seedling stage to 40 mmol/L of saline/alkali mixture and found that five of them were highly tolerant. Shi et al (2024) considered that 0.8% NaCl could be used for identificaiton of salt tolerance of green bean seedlings, based on the test of 68 Mungbean at germination stage. Based on the lab experiments, Hu et al. (2022) identified 24 high salt-tolerant mungbean lines by using 150 mmol NaCl and reported that the fresh or dry weight of upground/underground and the length and branches of root could be used as index for assessment of salt tolerance. Jiao et al. (1997) identified 180 and 16 tolerant accessions in germination and seedling stages from 1199 mungbean accessions using 1.8% NaCl and found that there is no obvious relationship of tolerance between germination and seedling stages. He (2021) analyzed the salt tolerance of 19 mungbean in germination stage by collecting 30 agronomic traits and found that 1% NaCl could be used for screening the salt tolerance in germination stage. Based on 50 mmol/L mixed alkali ($NaHCO_3$:Na2CO_3 with the mole ratio of 9:1), Xu et al. (2017) identified nine mungbean that were tolerant to alkali at germination stage. Wang et al. (2014) measured the drought tolerance of 70 mungbean based on repeat drought stress in green house and recommended that the survival rate of seedling under the first drought stress was an important index in early stage for screening a large number of germplasm resources of mungbean. Based on the survival rate of seedlings under drought stress in field test, Wang et al. (2015) identified that 16 mungbean lines were highly tolerant to drought and deduced a protocol for identifying and evaluating drought resistance in mungbean. They also assessed the germination ability of mungbean lines with PEG-6000 manipulation drought condition (Wang et al. 2014). Based on the germination rate of four accessions of mungbean germplasm resources with 5 PEG-6000 concentrations for drought resistance, Wang et al. (2014) deduced a set of index that could be used for the assessment of salt tolerance during the germination stage and after assessment of 21 mungbean accessions for drought resistance in the different irrigation conditions of greenhouse, they found that drought resistance index was recommended to screen a large number of germplasm resources of Mungbean at adult stage (Wang et al. 2015). By simulating drought condition, Zhao et al. (2017a) studied the drought resistance of different mungbean varieties and identified nine lines that were highly tolerant to drought during the whole growth period. Zhang et al. (2023) identified 3 from 40 mungbeans that were highly tolerant to drought at seedling stage using 15% PEG6000 treatment. Huang et al. (2021) identified 2 from 56 mungbeans that were highly tolerant to drought at germination stage using 10% mannitol solution to simulate drought stress. Based on 215 mungbean core germplasm resources and 85 new breeding lines, Shen et al. (2022) deduced that germplasm from northern part had high level of resistance to wild than those from southern part at the seedling stage. Liang et al. (2017) identified 3 of 61 mungbean lines that are tolerant to low temperature (10 °C) at germination stage, providing important reference for cultivation and breeding of mungbean in north regions. Based on the establishment of a protocol for wilt resistance, Zhu (2018) assessed the resistance of 433 mungbean lines and identified 54 lines are highly resistant to fusarium wilt. Singh et al. (2013) used a combination of conventional survey strategy, participatory varietal selection, molecular markers, and chemical analysis and revealed that the farmers had the capacity to utilize available genetic resources to manage disease and they can identify the disease at early stages of plant development. Aski et al. (2021) assessed the variations on fourteen root-related traits under modified semi-hydroponic screening conditions among 296 genotypes mini ore collections of mungbean, and identified a set of germplasm that could enhance adaptation to water-deficit stress and other stress conditions. Based on the germination rate of four accessions of mungbean germplasm resources with 5 PEG-6000 concentrations for drought resistance, Wang et al. (2014) deduced a set of index that could be used for assessment of salt tolerance during the germination stage.

1.3 Evaluation on Nutritional Quality

With the human recognition on healthy function of mungbean food, more and more research had put on the bioactive factors in mungbean. Based on the sensory evaluation of boiled beans, Zhao et al. (2022) analyzed 26 mungbean cultivars and selected some suitable ones for fresh food processing. After the assessment of the genetic architecture of micronutrients (grain iron and zinc concentration) and anti-nutritional factors (grain phytic acid and tannin contents) in association with mapping panel of 145 diverse mungbeans, Sinha et al. (2023) found that some genotypes had high grain iron and zinc concentrations but low grain phytic acid and tannin contents. Ullah et al. (2014) tested the contents of protein and crude fat of three mungbean cultivars and found that there were obvious difference among cultivers, indicating it is possible to screen mungbean lines with high nutrition contents for processing, further assessment suggested that aspartic and glutamic acids, were in considerable amount ranged from 13 to 23% followed by leucine, isoleucine, alanine, valine, lysine, phenyl alanine, serine, and arginine which fell in the range of 3–8% of total protein and there were difference of content among cultivars. In addition, He and Wang (1987) assessed the protein, amino acid, and starch contents of 700 mungbean accessions and found that the protein content was two to three times higher than corns and especially lysine. Lv et al. (2023) evaluated the production and quality of sprout of a set of elite germplasm and recommended that the edible fresh quality, soluble protein content, total phenol content, DPPH free radical scavenging rate, and hypocotyl length could be used as evaluation indexes for mungbean sprouts. Yuan et al. (2012) evaluated the sprout production of 470 accessions and found that there were differences in thickness, length, flavor, and production of sprouts among different mungbeans, and the indices of sprouts had no relationship with the content of nutritional factors; however, small seeds had a higher sprout production. Wang et al. (2024) analyzed the sprout production of 35 new mungbean lines and suggested that the thickness of hypocotyl of sprout could be an important index for screening Mungbean line for sprout production. Zhao (2023) assessed the content of vitexin/isovitexin in 376 mungbean, and screened a set of elite germpalsm with high content of vitexin/isovitexin.

1.4 Evaluation Using DNA Marker

Mungbean is sensitive to light and temperature, so phenotypes of germplasm resources always vary under different environments, so it is difficult to precisely differentiate a large set of germplasm. DNA markers are much stable in reflecting the genetic background of whole genome. However, the development of DNA markers for mungbean was much lagged than those major crops for a long period, because only common markers, such as RAPD, and AFLP, were usually used in the evaluation of mungbean collections in the past. Then, SSR markers from relative species-adzuki bean were tested for transferability and used for mungbean diversity analysis, but the markers available were much limited, leading to a low level of diversity analysis and gene mapping. With the development of DNA sequencing technology, DNA marker from mungbean itself was gradually identified and then widely used in the assessment of germpalsm resources and construction of genetic maps. For instance, the polymorphism of 44 mungbean genotypes was assessed based on nuclear ribosomal DNA and RAPD to provide useful information for genetic improvement program, and RMG-62, RMG-976, and NDM-56 were deemed to be potential source of parents for crop improvement (Raturi et al. 2012). Sequence-related amplified polymorphism (SRAP) was also used for the assessment of 21 mungbean from India, and the results indicated that this kind of markers were much efficient for the identification of mungbean (Aneja et al. 2013). SSR analysis among 157 cultivated and wild mungbean also showed that wild genotypes separated well from cultivated ones, but no obvious geographical distinctions existed between genotypes collected within China (Chen et al. 2015). In addition, 21 mungbeans from India were evaluated using 42 RAPD, ISSR, and SSR markers, and less genetic divergence among them was found (Tabasum et al. 2020). In Kenya, seven mungbeans were evaluated based on 13 traits and 10 SSR markers, and 18 significant associations were revealed by marker traits association analysis (Mwangi et al. 2021). There are also many reports on the genetic diversity analysis mung bean germplasm resources. For instance, based on the analysis of 352 introduced mungbean germplasm with SSR markers, Qiao et al. (2015) pointed out that the diversity of mungbean from East Asia was much lower than the Chinese mungbean germpalsm and that accessions from Russia and USA, indicating that accession from these countries were collected from diverse regions. While Zhao et al. (2017b) investigated the diversity of 156 mungbean from Northeast China using SSR and found that there were distinct differentiations among accession from different provinces. Recent studies based on SSR analysis suggested that mungbean within a province had a similar genetic relationship (Zhang et al. 2024; Zhao et al. 2018). Studies also showed that the breeding line had a narrower genetic background than landraces, which was a disadvantage for stable production of this crop (Wang et al. 2024). Raturi et al. (2012) evaluated the diversity of 44 mungbean based on RAPD.

2 Conclusion

UP to now, a large amount of mungbean germplasm has been collected worldwide and evaluated on agronomic traits that useful for production, resistance to disease or pests, tolerance to extreme climates and even nutrition fators to make a full use of superior resources and to achieve new variety selection is important direction for future effort.

References

Aneja B, Yadav NR, Yadav RC, Kumar R (2013) Sequence related amplified polymorphism (SRAP) analysis for genetic diversity and micronutrient content among gene pools in Mungbean [*Vigna radiata* (L.) Wilczek]. Physiol Mol Biol Plants 19(3):399–407. https://doi.org/10.1007/s12298-013-0177-3

Aski MS, Rai N, Reddy VRP, Gayacharan, Dikshit HK, Mishra GP, Singh D, Kumar A, Pandey R, Singh MP, Pratap A, Nair RM, Schafleitner R (2021) Assessment of root phenotypes in Mungbean mini-core collection (MMC) from the World Vegetable Center (AVRDC) Taiwan. PLoS One 16(3):e0247810. https://doi.org/10.1371/journal.pone.0247810

Bisht IS, Mahajan RK, Patel DP (1998) The use of characterization data to establish the Indian Mungbean core collection and assessment of genetic diversity. Genet Resour Crop Evol 45:127–133. https://doi.org/10.1023/A:1008670332570

Chang Y, Peng L, Ji l, Wang S, Wang L, Wu J (2023) Genome-wise association study identified genomic regions associated with drought tolerance in Mungbean (*Vigna radiata* (L.) R. Wilczek). Theor Appl Genet 136:40. https://doi.org/10.1007/s00122-023-04303-3

Chen H, Qiao L, Wang L, Wang S, Blair MW, Cheng X (2015) Assessment of genetic diversity and population structure of mung bean (*Vigna radiata*) germplasm using EST-based and genomic SSR markers. Gene 566(2):175–183. https://doi.org/10.1016/j.cj.2017.08.004

Diatta AA, Abaye O, Battaglia ML, Leme JF, Seleiman M, Babur E, Thomason WE (2024) Mungbean [*Vigna radiata* (L.) Wilczek] and its potential for crop diversification and sustainable food production in Sub-Saharan Africa: a review. Technol Agron 4(1):e031. https://doi.org/10.48130/tia-0024-0030

Gao Y (2018) Analysis and evaluation of agronomic traits of Mungbean germplasm resources. Hebei Agric Sci 22(03):70–74

Han X, Li L, Chen H, Liu L, Sun L, Wang X, Xiang Y, Wan Z, Liu C (2022) Re sequencing of 558 Chines eMungbean landraces identifies genetic loci associated with key agronomic traits. Front Plant Sci 13:1043784. https://doi.org/10.3389/fpls.2022.1043784

Hanumantha Rao B, Nair RM, Nayyar H (2016) Salinity and high temperature tolerance in Mungbean [*Vigna radiata* (L.) Wilczek] from a physiological perspective. Front Plant Sci 7:957. https://doi.org/10.3389/fpls.2016.00957

He R (2021) Study on genetic diversity of Mungbean and evaluation of salt tolerance during seed germination. Inner Mongolia Agricultural University

He W, Wang W (1987) Preliminary study on the nutritional quality identification of Mungbean germplasm resources in China. Acta Agron Sin 4:346–348

Hou XF, Wang CP, Liu J, Song RJ, Cheng YH, Li T (2020) The correlation and cluster analysis of main agronomic traits of Mungbean. Agric Sci Technol Commun 2:134–140

Hu L, Wang S, Wang L, Cheng X, Chen H (2022) Identification of salt tolerance of Mungbean germplasm resources at the seedling stage and screening of salt-tolerant germplasm. Acta Agron Sin 48(02):367–379. https://doi.org/10.3724/SP.J.1006.2022.04283

Huang N, Zhu Z, Liu C, Liao F, Li L, Chen H, Liu L, Han X, Wan Z, Sha A (2021) Screening and evaluation of drought-tolerant germplasm resources during the germination period and analysis of genetic diversity of Mungbean. Fujian J Agric Sci 36(03):255–263. https://doi.org/10.19303/j.issn.1008-0384.2021.03.002

Jiao G, Ren J, Lu G, Jiang C (1997) Identification and research on salt tolerance of Mungbean variety resources. Crop Var Resour 2:39–41

Kyu KL, Taylor CM, Douglas CA, Malik AI, Colmer TD, Siddique KHM, Erskine W (2024) Genetic diversity and candidate genes for transient water logging tolerance in Mungbean at the germination and seedling stages. Front Plant Sci 15:1297096. https://doi.org/10.3389/fpls.2024.1297096

Lang B, Wang W, Feng R, Zhou Z, Chen Y, Zhao Y, Jin F, Du Y (2024) Comprehensive evaluation of salt-alkali tolerance and analysis of salt-alkali tolerance mechanism of 35 Mungbean germplasm resources during germination period. Shandong Agric Sci 56(05):69–79

Li L, Liu C, Chen H, Wan Z, Liu L, Wu G, Li D (2009) Identification of local germplasm resources of Mungbean in Hubei and evaluation of their main agronomic traits. Agric Sci Technol 18(2):2219–2228

Liang X, Qin P, Yan M, Leng D, Zhang W, Liu C, Fang S, Zheng D (2017) Screening of low temperature tolerant Mungbean germplasmresources. J North Agric 45(01):18–24

Liu C, Wang S, Wang L, Sun L, Mei L, Xu N, Cheng X (2008) Establishment of candidate core collection in Chinese Mungbean germplasm resources. Acta Agron Sin 34(4):700–705. (in Chinese)

Liu X, Xing B, Wu R et al (2019) Variation of main agronomic traits of Mungbean in northern Shanxi and its impact on yield composition. Crop J 05:69–75

Liu C, Wang Y, Peng J, Fan B, Xu D, Wu J, Cao Z, Gao Y, Wang X, Li S, Su Q (2022) High quality genome assembly and pan-genome studies facilitate genetic discovery in Mungbean and its improvement. Plant Commun 3(6):1–16. https://doi.org/10.1016/j.xplc.2022.100352

Lu H, Zeng L, Wang C et al (2021) Identification and evaluation of 28 Mungbean germplasm resources in western Heilongjiang Province. Seed Sci Technol 39(19):3–6

Luo G, Lin S, Chen Y, Cai Q (2010) A comprehensive analysis of major agronomic traits of Guangxi Mungbean germplasm. Guangxi Agric Sci 4112:1259–1261

Lv C, Zhang X, Huang L, Xue C, Yuan X, Chen X (2023) Evaluation of sprout characteristics of different Mungbean varieties and screening of special varieties. J Jiangsu Agric Sci 51(04):152–163. https://doi.org/10.3969/j.issn.1004-1524.2023.01.11

Mwangi JW, Okoth OR, Kariuki MP, Piero NM (2021) Genetic and phenotypic diversity of selected Kenyan Mungbean (*Vigna radiata* L. Wilckzek) genotypes. J Genet Eng Biotechnol 19(1):142. https://doi.org/10.1186/s43141-021-00245-9

Qiao L, Chen HL, Wang LX, Wang SH, Cheng XZ, Zhang YW (2015) Genetic diversity of foreign Mungbean germplasm resources by agronomic characters. J Plant Genet Resour 16(5):986–993

Raturi A, Singh SK, Sharma V, Pathak R (2012) Molecular characterization of *Vigna radiata* (L.) Wilczek genotypes based on nuclear ribosomal DNA and RAPD polymorphism. Mol Biol Rep 39(3):2455–2465. https://doi.org/10.1007/s11033-011-0996-7

Schafleitner R, Nair RM, Rathore A, Wang Y, Lin C, Chu S, Lin P, Chang J, Ebert AW (2015) The AVRDC - the World Vegetable Center Mungbean (*Vigna radiata*) core and mini core collections. BMC Genomics 16(1):344. https://doi.org/10.1186/s12864-015-1556-7

Shen Y, Zhang Z, Sun S, Wang Y, Fan B, Liu C, Wang S, Su Q, Shi H, Zhu Z, Tian J (2022) Identification of *Fusarium* wilt resistance of Mungbean germplasm resources and screening of resistance resources. J Plant Genet Resour 23(06):1660–1669. https://doi.org/10.3969/j.issn.1002-2481.2022.07.05

Shi H, Fan B, Liu C, Wang Y, Wang S, Zhang Z, Su Q, Tian J (2024) Identification and evaluation of salt tolerance of Mungbean germplasm resources during germination period. Crop J:1–13

Singh R, van Heusden AW, Kumar R, Visser RG, Yadav RC (2013) Genetic diversity of Mungbean (*Vigna radiata* L.) in iron and zinc content as impacted by farmers' varietal selection in Northern India. Ecol Food Nutr 52(2):148–162. https://doi.org/10.1080/03670244.2012.706006

Sinha MK, Aski MS, Mishra GP, Kumar MBA, Yadav PS, Tokas JP, Gupta S, Pratap A, Kumar S, Nair RM, Schafleitner R, Dikshit HK (2023) Genome wide association analysis for grain micronutrients and anti-nutritional traits in Mungbean [*Vigna radiata* (L.) R. Wilczek] using SNP markers. Front Nutr 10:1099004. https://doi.org/10.3389/fnut.2023.1099004

Sokolkova A, Burlyaeva M, Valiannikova T, Vishnyakova M, Schafleitner R, Lee C, Ting C, Nair RM, Nuzhdin S, Samsonova M, Wettberg E (2020) Genome-wide association study in accessions of the mini-core collection of Mungbean (*Vigna radiata*) from the World Vegetable Gene Bank (Taiwan). BMC Plant Biol 20(Suppl 1):363. https://doi.org/10.1186/s12870-020-02579-x

Tabasum A, Hameed A, Asghar MJ (2020) Exploring the genetic divergence in Mungbean (*Vigna radiata* L.) germplasm using multiple molecular marker systems. Mol Biotechnol 62(11–12):47–556. https://doi.org/10.1007/s12033-020-00270-y

Ullah R, Ullah Z, Al-Deyab SS, Adnan M, Tariq A (2014) Nutritional assessment and antioxidant activities of different varieties of *Vigna radiata*. Sci World J 1:871753

Wang L, Wu J, Jing R, Cheng X, Wang S (2014) Identification of drought resistance of Mungbean germplasm resources at the bud stage. J Plant Genet Resour 15(03):498–503

Wang L, Wu J, Jing R et al (2015) Identification of drought resistance of Mungbean germplasm resources at the seedling stage. Acta Agron Sin 41(01):145–153

Wang S, Gong D, Liu H, Pan X, Wang S, Li Z, Wang L (2024) Analysis of bud characteristics of innovative Mungbean germplasm. J Plant Genet Resour 25(01):141–146

Wu X, Islam ASMF, Limpot N, Mackasmiel L, Mierzwa J, Cortés AJ, Blair MW (2020) Genome-wide SNP identification and association mapping for seed mineral concentration in Mungbean (*Vigna radiata* L.). Front Genet 11:656. https://doi.org/10.3389/fgene.2020.00656

Xu N, Chen B, Wang M, Bao S, Wang G, Guo Z (2017) Identification of alkaline tolerance of Mungbean varieties during germination period. Acta Agron Sin 43(01):112–121

Yuan X, Zhang T, Cheng X, Wang L, Wang S (2012) Evaluation and screening of sprout characteristics of Mungbean germplasm resources in my country. J Plant Genet Resour 13(05):879–883
Zhang Y, Zhang Y, Zhang Z (2013) Analysis of phenotypic traits diversity of Mungbean germplasm resources. Acta Agron Sin 3(01):15–19
Zhang S, Guo Y, Ai J, Zhang P, Wang F (2023) Screening of drought resistance indexes and identification of drought resistance of Mungbean resources during germination period. Shanxi Agric Sci 51(02):143–150
Zhang Y, Ding D, Li M, Cai D, Liu J, Cao L, Chen J, Yang S (2024) SSR analysis of genetic diversity of local Mungbean germplasm resources in Henan Province. Mol Plant Breed:1–8
Zhao Y (2023) Screening of high vitexin Mungbean germplasm resources and study on metabolic regulation during grain development. South China University of Technology
Zhao X, Zhang Z, Zhu H et al (2017a) Identification of drought resistance and morphological indicators of Mungbean during the whole growth period. Acta Agric Boreali-Sin 32(S1):180–184
Zhao Y, Wang Y, Zhang D, Wang L, Zuo Z (2017b) SSR analysis of genetic diversity of Mungbean germplasm resources in Northeast China. J Nucl Agric Sci 31(12):2340–2349
Zhao Y, Wang Y, Zhang D (2018) Construction of SSR fingerprints and variety identification of Mungbean varieties in Liaoning region. Food Sci 39(06):284–290
Zhao T, Meng X, Chen C, Wang L, Cheng X, Xue W (2022) Agronomic traits, fresh food processing characteristics and sensory quality of 26 Mungbean (*Vignara diata* L.) cultivars (Fabaceae) in China. Foods 11(12):1687
Zhu L (2018) Identification of races of the pathogen causing Mungbean wilt and screening of resistance sources. Chinese Academy of Agricultural Sciences, pp 18–22

[illegible] Zhang L, Chen X, [illegible] Shandong [illegible]

[illegible] Y, Zhang [illegible] Y, Zhang [illegible]

[illegible] Plant [illegible]

[illegible] Sci [illegible]

[illegible] North Plain [illegible]

[illegible] South [illegible]

[illegible] Y, Zhang Z, Zhao [illegible]

[illegible] Y, Zhang [illegible]

Status and Perspectives of Mungbean Breeding

Jing Tian and Baojie Fan

Abstract

Mungbean, also known as green gram, is widely grown in temperate, subtropical, and tropical regions. Owing to its nitrogen fixing, rich in nutritional substances, and long suitable sowing period, mungbean plays many valuable roles in soil fertility improvement, human dietary enrichment, cropping systems, and the processing of agricultural products. So, it is a crop with great potential to development in the world. New plant variety is an important component for agriculture. Mungbean variety development, popularization, and promotion greatly rely on the public sector, especially the international collaboration. In this chapter, we introduced the status of mungbean breeding in major countries like India, Myanmar, China, Thailand, the Philippines, Australia, Kenya, the Republic of Korea, and Pakistan. We also discussed the challenges and perspectives of mungbean breeding. We think that the major breeding objectives of mungbean should include high and stable yield, ideal plant types for cropping system and mechanical harvest, resistance to multiple diseases, tolerance to abiotic stresses, good seed quality for market demands, and food processing. The intra-specific cross will still be the main means of mungbean breeding in the future. And with the development of available molecular markers, molecular marker-assisted selection (MAS) would be the powerful method for mungbean breeding in the near future.

Keywords

Mungbean · Breeding · Status · Perspectives

J. Tian (✉) · B. Fan
Institute of Cereal and Oil Crops, Hebei Academy of Agricultural and Forestry Sciences, Shijiazhuang, Hebei, China

S. Paramasivam et al. (eds.), *Mungbean*,
https://doi.org/10.1007/978-981-95-7637-1_2

1 History of Mungbean Breeding

Mungbean (*Vigna radiata* L.), also known as green gram, is extensively cultivated across temperate, subtropical, and tropical regions globally, with a production area of 7.3 million hectares and an average yield of 721 kg/ha (Nair and Schreinemachers 2020). The production of this crop has historically been primarily located in Asia (Nair and Schreinemachers 2020; Anonymous 2021), including India, Myanmar, China, Thailand, Indonesia, Pakistan, the Philippines, Sri Lanka, Cambodia, Vietnam, Nepal, and Bangladesh, but has greatly increased in recent years in non-Asian countries such as Australia, Ethiopia, Uzbekistan, Venezuela, Kenya, South Africa, and Brazil. Owing to its long sowing window, nitrogen-fixing ability, short growth cycle, and adaptability to barren soils, mungbean serves as an ideal rotation crop in cereal-based systems and is valuable for disaster recovery and wasteland reclamation (Yimram et al. 2009). Nutritionally, the seeds are rich in protein (22–26%), dietary fiber (3–4%), vitamin B2 (0.25–0.29 mg/100 g), calcium (80–155 mg/100 g), and phosphorus (220–417 mg/100 g). They also contain functional compounds like vitexin, isovitexin, and D-chiro-inositol, which confer health benefits such as heat-clearing, detoxification, and protection for the liver and vision (Tang et al. 2014; Cao et al. 2011). As a result of these attributes, mungbean contributes significantly to improving soil structure, enriching human dietary, and fostering the processing and utilization of agricultural commodities.

The development of new plant varieties is a crucial component of agricultural progress. However, the private seed sector has limited interest in mungbean due to the absence of commercial hybrids and the ease with which farmers can save their own seeds. Consequently, variety development and scaling efforts rely primarily on the public sector. Mungbean improvement research is carried out by national agricultural research systems in Bangladesh, India, Myanmar, Pakistan, and other Asian countries including China, Thailand, Vietnam, and the Philippines, as well as in Australia. International collaboration has always been a key feature of mungbean improvement programs across Asia. Notably, the World Vegetable Center has played a vital role since 1973 by connecting national programs, sharing germplasm, and testing advanced breeding lines across different countries, initiating an international mungbean breeding program (Fernandez and Shanmugasundaram 1988).

A notable success was achieved in the early 1980s, when newly developed mungbean lines—derived from crosses between climate-resilient and disease-resistant germplasm from India and high-yielding, early, and uniformly maturing germplasm from the Philippines—were evaluated in international mungbean trials. Among them, two lines, VC 1973A and VC 2778A, were identified as especially promising, which were subsequently released as KPS 1 and KPS 2 in Thailand, and as Zhong Lu #1 and Elu #2 in China, eventually becoming dominant varieties in these two countries as well as several others (Shanmugasundaram et al. 2009). In the 1990s, the World Veg mungbean breeding program broadened its focus to incorporate resistance to mungbean yellow mosaic disease (MYMD), which had emerged as a major constraint to mungbean production across large areas of South Asia and Myanmar. In a collaborative effort, the Nuclear Institute for Agriculture and Biology

(NIAB) and the World Vegetable Center crossed VC1973A with a resistant line developed by irradiating a local variety. After several generations, this work yielded two advanced MYMD-resistant lines, NM92 and NM94 (Ali et al. 1997). NM94 has since become the most popular mungbean variety in Bangladesh (where it is known as BARI Mung-6) and Myanmar (known as Yezin-11), and the second-most popular variety in India (known as SML 668).

By now, the international mungbean breeding program at the World Vegetable Center had become a major source of improved varieties in Southeast Asia. Survey data indicated that 64% of the total mungbean area across 13 countries was planted with improved varieties, of which 32% were directly derived from the World Vegetable Center. According to the survey by Pepijn Schreinemachers et al., the adoption rates of varieties containing improved germplasm developed by the World Vegetable Center in collaboration with national partners were particularly high, accounting for 93% of the mungbean area in Pakistan, 77% in Myanmar, 67% in Bangladesh, and 27% in India (Schreinemachers et al. 2019).

2 Status of Mungbean Breeding in Major Countries

According to the growing areas, the major countries of mungbean production include India, Myanmar, China, Kenya, Tanzania, Indonesia, Pakistan, Bangladesh, Vietnam, Thailand, the Philippines, Sri Lanka, Australia, Ethiopia, Uzbekistan, etc. So the breeding of new variety in these countries plays an important role in the global mungbean production.

2.1 Mungbean Breeding in India

India is the largest producer and consumer of mungbean in the world, with a cultivated area of approximately 4.5 million hectares and an annual production of 2.5 million tons (Anonymous 2021). The systematic germplasm collection and evaluation were back to the 1920s, leading to the isolation of pure lines (Bose 1932). The first mungbean cultivar made available to farmers, "Type 1," was released in 1936. This was followed by "Type 44" in 1948, which marked the first cultivar developed specifically through hybridization and selection for agricultural use. But before 1960s, the breeding methods were mostly introduction or selections from landraces. And most varieties exhibited small-seeded, twining, prostrate, pod-shattering characteristics, longer maturity periods (>90 days), photoperiod sensitivity, and high susceptibility to yellow mosaic disease, as well as low inherent yielding ability (only 0.4–0.8 ton/ha).

During 1970–2000, biparental crosses and mutations became the major selection methods. The new varieties had better traits, such as semi-erect, non-twining, mostly tolerant to yellow mosaic disease and resistant to CLS (*Cercospora* leaf spot) and PM (powdery mildew), and improved inherent yielding ability. The landmark varieties are PDM 139, PDM 11, PDM 54, ADT 3, Co 5, Narendra Mung 1, LGG460, Co6, Pusa 9531, TARM 1, ML1613, MUM2.

From 2000 until now, the new varieties were mostly selected from biparental and multi-parental crosses and inter-specific crosses, which owned more comprehensive characters such as erect, non-twining, non-shattering, synchronous types, early maturing (60–65 days), highly resistant to yellow mosaic disease, resistant to CLS and powdery mildew, high inherent yielding ability. Such varieties are IPM 205–7 (Virat), IPM 410-3 (Shiha), Soorya (IPM 512-1), MH 421, IPM 2-14, IPM 02-3, Pusa 0672, HUM 12, HUM 16, HUM 2, IPM 99-125, MH318, SML668, and DGGV-2.

From 1985, a total of 143 varieties were released, including 59 national and 84 state. 76 varieties were in the seed chain, top 10 varieties of which shared >90% of total seed indent. The most widely used mungbean varieties were Shikha and Virat, accounting for 28.75% and 19.55% in national seed indent, respectively, in 2024.

There are two main cultivation types of mungbean in India, single cropping in spring/summer and rice fallow. The objectives of mungbean breeding in the spring/summer season are as follows: shorter duration (55–60 days), medium plant height (60–80 cm), determinate growth habit and synchronous maturity, high initial growth vigor and deep root system, more number of pods at top of plant and non-shattering habit, longer pods with >10 seeds/pod, and tolerance to terminal heat stress. For the rice fallow regions, the varieties should have such ideal plant types as follows: prolific root system and narrow leaf, early seeding vigor and high biomass, photo-thermo-insensitivity, early and synchronous maturity (55–60 days), shattering resistance, terminal drought tolerance, and multiple disease resistance. Both cultivation types need restructuring plant types for high density planting and mechanical harvesting by pre-breeding and alien gene introgression. Their productivity target is to enhance yield up to 2.0–2.5 t/ha.

2.2 Mungbean Breeding in Myanmar

Myanmar is the second-highest producer and largest exporter, with about 1.2 million ha cultivated and a total production of 1.5 million tons (MAOLI 2019). In Myanmar, mungbean is the main economic crop with 0.69 million ton exported volume, 1.2 million ha cultivated area, 1.55 million ton production, and 1449 kg/ha productivity. It is a short-duration grain legume and can be grown during monsoon (May–June), mid-monsoon (July–August), and pre-monsoon (February–March) with irrigation in the central dry zone. In the Delta region during the post-monsoon (November), under the field after rice-harvesting, Mungbean yellow mosaic disease (MYMD) is the most important constraint on yield in mungbean accounting for about 50%–100% of the yield reduction. Hot weather with little or no rainfall was conducive to the development of disease. In addition to MYMD, the incidence of dry root rot (DRR) disease increases during the crop period because of less and irregular rainfall and rising temperature (Ministry of Agriculture 2016; Shanmugasundaram et al. 2009).

So, the target variety profile for mungbean is as follows: high yield (>10%); early and synchronized maturity (60–65 days); high resistance to MYMD, resistance and moderately resistance to dry root rot disease and *Cercospora* leaf spot disease; resistance to bruchid; export quality of shiny green seed and sprout (100-seed weight > 6.5 g); heat tolerance (>35 °C); and non-shattering.

By now, DAR have released 19 improved varieties, 12 of which came from World Vegetable Center (Schreinemachers et al. 2019). The main varieties widely popularized in Myanmar are Yezin 9 (VC 1973A), Yezin 11 (NM94), Yezin 14 and MAS 1 which show high yield, early and uniform maturity, resistance to MYMD (Yezin 11 and Yezin 14). Now, the improved varieties are planted on 89% of the mungbean area in Myanmar.

Yezin 16 was released in 2022. AVMU 1688 (Yezin 17) is under the registration process for varietal release. By crossing of market preferred varieties (Yezin 11, Yezin 14, Yezin 9) with MYMV-resistant accession (VI 004734AG) and with DRR-resistant accession (VI 001482BG), promising lines with MYMD and DRR-resistant genes pyramided were successfully developed. Among them, YM-22-2-22 and YM-22-2-9 showed high yield, synchronized maturity for mechanical harvest, large seeded export for sprout markets.

2.3 Mungbean Breeding in China

In China, the improvement of mungbean varieties can be traced back to the beginning of 1980s, when Lufeng No. 1 and Lufeng No. 2 were bred out by pedigree selection from the local varieties Lindian mungbean and Nehe mungbean by Heilongjiang Academy of Agricultural Sciences.

The genuine breeding and improvement of mungbean began in the mid-1980s. Based on the collection, identification, and catalog of germplasm resources from 1980 to 1985, most academies and institutes carried out the purification and rejuvenation, introduction of excellent varieties, and systematic breeding of local varieties. By the end of the twentieth century, a total of 22 new mungbean varieties had been developed, with pedigree breeding accounting for 72.72% of the selection methods. Among these, Zhonglv No. 1 (VC1973A) emerged as the first major variety to be introduced and widely adopted throughout China.

Since the start of the twenty-first century, a growing number of research institutes have engaged in mungbean crossbreeding. By 2007, 34 new varieties had been developed, with hybridization accounting for 61.76% of the breeding methods used. In this period, varieties such as Zhonglu No. 2, Jilu No. 2, Jilu No. 7, Baolu No. 942, Bailu No. 8, Bailu 522, Yulu No. 4, Weilu No. 4, and Jinludou No. 1 etc., played an important role in the mungbean production. The breeding of these new varieties initially solved the problems of long growth period, creeping, lodging, low yield, and poor adaptability at that time in China (Tian et al. 2021).

The inception of the Chinese Agricultural Research System-Food Legumes program in 2008 acted as a catalyst for mungbean breeding. By 2022, 98 new varieties had been registered. The breeding methodologies were dominated by crossbreeding, which produced 74 varieties (75.5%), while radiation mutagenesis yielded 7 (7.14%). These varieties such as Zhonglv 6, Zhonglv 7, Jilv 15, Jilv 17, Sulv 5, Sulv 6, and Jinlvdou 7, which are resistant to bruchids, have solved the problem of bruchid damage during storage. Concurrently, adaptations for mechanization were achieved in varieties like Jilv 19 and Zhonglv 12, which exhibit a compact architecture, concentrated pod set, and uniform maturity traits that collectively enable efficient mechanical harvesting.

By statistics from the Chinese Agricultural Research System-Food Legumes, the yield of the new varieties was increased by 12.0%–38.6% compared with that of the local varieties. Through the extension and application, more than 60% of the main varieties have been replaced by new varieties in China. At present, the main bred varieties of mungbean in China are the series of ZhongLu, Jilu, Bailu, Jiilu, Jinlu, and Sulu, etc.; especially, Zhonglu No. 4, Zhonglu No. 5, Jilu No. 7, Jilu No. 13, Bailu No. 6, Bailu No. 8, Jinludou No. 6, Sulu No. 2, and Sulu No. 5 have become the major varieties in China (Tian et al. 2021).

Despite these advances, mungbean varieties in China still have various deficiencies as follows:

1. It is necessary to the develop new varieties suitable for mechanized harvest.

   At the beginning of this century, because of the small planting scale and low labor costs, the cultivation of mungbean was mostly done by artificial sowing, management, and harvest. In recent years, along with the increase of labor cost, the benefit of mungbean planting was decreased significantly. Mechanized production, especially mechanized harvest, has become the key factor to increase the expansion of planting scale and planting benefit of mungbean.
2. To develop new varieties with multiple resistance to major diseases.

   In recent years, the main diseases that affected mungbeans in China are bacterial blight, leaf spot, *Fusarium* wilt, powdery mildew, and virus. Among them, *Fusarium* wilt was prevalent in nearly all production areas and caused more and more serious damage. Bacterial blight is an important disease in north, northeast, and northwest of China. Because of the change of the epidemic diseases, the resistant varieties used in production have become susceptible to other disease. For example, Zhonglu No. 5 was resistant to leaf spot but susceptible to *Fusarium* wilt. Jilu 7 showed good resistance to virus but poor resistance to bacterial blight.
3. Lack of special varieties for food processing

   Mungbean is a kind of crops with high-protein and low-fat contents, and has medical and food double use. The main season of consumption in China is summer. Mungbean soup has the capability to refresh the mind, alleviate heat stroke, and reduce swelling. And bean sprout and noodle are the main products of Mungbean processing at present. But now, there is a lack of special varieties suitable for bean sprout and noodle processing. There is even a shortage of the varieties with high protein content, high vitex and isovitex content, high iron content, and special effects, such as strong heat-detoxifying and hypoglycemic effect.

The objectives of mungbean breeding in China must be adapted to the needs of agricultural production and marketing. So, the specific breeding objectives for future are as follows: (1) Resistance to disease and tolerance to abiotic adversity, such as bacterial blight, leaf spot, *Fusarium* wilt, powdery mildew and virus diseases, and salt soil; (2) High nutrition and health function varieties and processing-specific ones, such as those suitable for processing bean paste, bean sprout, protein isolates, protein concentrates, with high contents of protein, micronutrient (iron), special physiological active substances; (3) Characteristics suitable for mechanized production, such as early maturity, erect and compact plant, lodging resistance, high pod-bearing position, uniform maturity, non shattering of pod, etc.

Crossbreeding remains the predominant mungbean breeding method in China, yet it is encumbered by significant disadvantages. These include a protracted development timeline (often approximating 10 years per variety), low efficiency, and the challenge of efficiently pyramiding optimal traits. Molecular marker-assisted breeding (MAS) presents a transformative alternative, with the potential to markedly shorten breeding cycles and improve precision and efficiency. The continued development and identification of molecular markers associated with agronomically important traits—such as disease and insect resistance, drought tolerance, yield, and quality—are paving the way for MAS to become a central methodology in future mungbean improvement.

2.4 Mungbean Breeding in Thailand

Mungbean was first introduced in Thailand between the third and the first century B.C. According to the records of history books, the first cultivar used in Thailand in the 1930s was "Sri Samrong," a small-seeded type, probably less than half the size of the current cultivars which vary between 6.5 and 7.0 g/100 seeds. This cultivar was so sensitive to photo-periods that its planting data were fixed from September 15 to 30 each year, and an early rainy season planting often delayed its flowering for several months. From 1968 to 1969, some new lines introduced from Australia, India, and the Philippines were tested in Thailand. A new line M7A from the Philippines was well-adapted to the climatic conditions of Thailand, out-yielded many introduced lines, and had a more uniform maturity and large seed size, ca.6.5 g/100 seeds. So it was released in 1976 by the Thai Department of Agriculture (the DA) as U-thong 1 and has become very popular ever since. In 1985, two local cultivars "Maledman Phak Dam" and "Maledman Phak Khao" were also recommended to the farmers. All these cultivars had contributed significantly to the expansion of mungbean production in Thailand (Fernandez and Shanmugasundaram 1988).

Mungbean hybridization programs in Thailand began in 1978. Both the DA and Kasetsart University (KU) have identified several superior genotypes from the Asian Vegetable and Development Center (AVRDC)-improved mungbean lines, segregating materials, and elite lines of the IMNs coordinated by AVRDC. Consequent to numerous testings in cooperation with the DA, the UK, the Thai Department of

Agricultural Extension (DAE), and other colleges and centers released two mungbean varieties to the farmers in 1986: Kamphaengsaen 1 (VC 1973A) and Kamphaengsaen 2 (VC 2778A). Chai Nat 60 (VC 1178), obtained from the MG 50-10A (Y) X ML 6 cross, was released in 1988 by mungbean researchers at Chai Nat Field Crops Research Center, which showed short maturity (55 days, one week earlier than U-thong 1), good plant type, pod extrusion high above the canopy, large seed size, and high yield potential (yielded 20% over Uthong 1) under a higher plant population density. This variety was suitable for rice-based cropping systems in the early rainy season or after field rice in the late rainy season. The native variety Fug Koa, meaning white pod, was not a recommended variety but was very popular with farmers in the north and northeast, and although relatively late maturing compared to Uthong 1, it has a large seed size. Large-seeded mungbeans are preferred by the farmers, consumers, and processors.

Historically, all commercially available mungbean cultivars in Thailand have been characterized by medium seed size. This landscape has recently changed with the official registration of two new cultivars: 'KUML4' (Reg. no. CV-354, PI 703288) and 'KUML8' (Reg. no. CV-355, PI 703289). Developed through a combination of pedigree and bulk-selection methods, these new lines represent a significant advancement. In standard yield trials, KUML4 demonstrated high productivity with seed yields of 1444.75 kg/ha and 1316.38 kg/ha. A defining feature of both cultivars is their larger seed size. The 100-seed weight for KUML4 was 7.25 g and for KUML8 7.84 g, substantially exceeding the average of 6.58 g observed in the check cultivars (Somta et al. 2024).

In Thailand, mutation breeding was also used for developing high-yielding varieties. Mutant variety Chai Nat 72 is popular in the country for its higher yield, larger seed size, and better resistance to alkaline soil than the wild-type variety. Now molecular-assisted selection (MAS) has been used in the mungbean breeding. By using the bruchid-resistant varieties V4718 and V2802, as well as the sensitive variety CN84-1, as parents, a new bruchid-resistant and high-yielding strain was obtained through hybridization, multi-generation backcrossing, and multiple crossing. Molecular markers VrTAF5-Indel3 and DMB-SSR158 were used to assist in the selection of bruchid resistance (Yundaeng et al. 2021). At the same time, using one set of molecular markers VrML012-2, VrML012-7, PvM13, VrML012_indel3 and another set of molecular markers VrBrSSR-037, VrBr017, VR1D1, VR1D5b, molecular-marker assisted selection was performed on the populations derived from the crossing and back-crossing among varieties V4718, V2802, CN84-1, and V2709 for powdery mildew and bruchid resistance traits. A new strain resistant to both powdery mildew and bruchid was developed (Laosatit et al. 2024).

2.5 Mungbean Breeding in the Philippines

Over the past five years, mungbean production in the Philippines has shown a steady upward trend. In the third quarter of 2022, output of mungbean reached an estimated 1.37 thousand metric tons (MT). Cagayan Valley stands as the leading production region, with the Davao Region following as the second-largest contributor. Known locally as "mongo," mungbean is commonly intercropped or rotated with rice across the country. For instance, farmers in Cuyapo, Nueva Ecija, frequently cultivate mungbean after the first rice cropping season, taking advantage of the crop's tolerance to dry conditions (Domingo 2023).

The history of mungbean improvement in the Philippines dates back to before 1916, when various landraces and introduced germplasm underwent initial evaluation. Formal breeding programs commenced in 1933 (Catipon et al. 1988). Early efforts involved the collection of landraces, from which pure lines were isolated, selected, and evaluated. This work led to the release of six popular pure-line cultivars to farmers (Ballon et al. 1977).

A national breeding program was established in 1956 with the objective of developing high-yielding cultivars through hybridization. The target traits included a semi-dwarf stature, uniform maturity, large and shiny seeds, non-shattering pods, and high protein and carbohydrate content. This initiative resulted in the 1969 release of the cultivars 'MG50-10A' and 'MD15–2' (Catipon et al. 1988). Yielding approximately 1.2 tons per hectare—double that of the earlier pure lines—these cultivars revolutionized yield potential, plant architecture, and phenology, though they remained susceptible to various diseases.

By the mid-1970s, utilizing resistant germplasm from India, breeders successfully developed cultivars such as CES 55, CES 87, and Pag-asa 1 (CES ID-21), which combined good yield, improved architecture, and enhanced disease resistance (Catipon et al. 1988). The 1980s saw the release of subsequent series, including the Pag-asa and BPI Mg lines.

Current breeding objectives focus on incorporating resistance to diseases like *Cercospora* leaf spot and powdery mildew, tolerance to insect pests, adaptability to low soil pH and moisture stress, and high nitrogen-fixing ability into high-yielding genetic backgrounds.

2.6 Munbgean Breeding in Australia

Mungbean is a tropical grain legume cultivated primarily in the summer-rainfall regions of northern Australia. Historically as an important export commodity, the crop yielded approximately 35,000 tonnes annually until 2009, with export earnings exceeding AU$25 million from markets including Japan, India, the United States, and the Philippines (AMA 2014). In recent years, production has expanded significantly, with the average annual output from 2018 to 2022 surpassing 90,000 tonnes. About 90% of this production is exported. Unlike many bulk grain crops, mungbean is marketed mainly as a high-value vegetable product, making seed appearance a critical quality attribute. Only a small proportion of the Australian harvest is used domestically, primarily for sprouting.

Mungbean cultivation in Australia began in the late 1960s and 1970s with the introduction of initial commercial varieties from the World Vegetable Center (AVRDC) in Taiwan, China. Subsequently, the Grains Research and Development

Corporation (GRDC) provided support for CSIRO to initiate a targeted breeding program suited to Australian growing conditions. The release of well-adapted varieties, such as 'Emerald,' played a key role in the industry's expansion, driving production from approximately 10,000 tonnes in the late 1980s to 20,000 tonnes by 1996, and further to around 45,000 tonnes by the mid-2000s. Following the release of the variety 'White Gold™,' CSIRO divested its investment in mungbean research in 2002, bringing the breeding program to a close.

In 2003, the Queensland Department of Agriculture, Fisheries and Forestry (DAFF), in partnership with the Grains Research and Development Corporation (GRDC), established the National Mungbean Improvement Program (NMIP) to enhance the productivity of the Australian mungbean industry. The program aimed to boost yield and production stability by leveraging genetic resources, improving regional adaptation, and introducing resistance to foliar diseases. This was achieved through extensive testing of advanced breeding lines across the northern grain-growing region. The release of the cultivars 'Crystal' and 'Satin' in 2008 significantly boosted grower and industry confidence. This was reflected in a notable expansion of the planting area from 45,000 to 66,000 hectares and an increase in production to 65,000 tonnes.

In 2021, an End Point Royalty (EPR) system was introduced to more effectively support ongoing varietal improvement and share production risks with growers. Under this system, a fee is levied on growers and directed to both breeders and their commercialization partners. The EPR mechanism provides breeders with crucial financial resources and market feedback, which in turn helps guide the development of new varieties that are better aligned with the practical needs of growers and the demands of end-user markets.

Now the varieties planted in Australia include Berken, CeleraII-Au, Crystal, Green Diamond, ade-AU, and Satin II. Berken, an old Oklahoman cultivar, with medium-large, evenly sized, bright-green seed, remains the preferred variety for sprouting, primarily because it produces large sprouts that are easy to market and premiums for sprouting grades.

However, the variety Berken is susceptible to three major diseases—tan spot, powdery mildew, and halo blight—and is also prone to lodging, weather damage, and pod cracking. In 2008, the dull-seeded variety Satin II was released to serve a niche market. Compared to its predecessor Satin, Satin II provides a significant yield increase of approximately 20%, along with improved resistance to tan spot and powdery mildew.

Celera II-AU, released in 2014 as a successor to Celera and Green Diamond, is a small-seeded variety characterized by a low proportion of hard seed. It represents the first mungbean cultivar in Australia with improved, moderate resistance to halo blight.

Green Diamond, released in 1997, is also a small-seeded type but exhibits a very high hard seed level (70%). It is relatively early-maturing, with an erect growth habit that holds pods high in the canopy. This variety often outperforms others under dry growing conditions.

Reselected Emerald, released in 1993, features medium-large, bright-green seeds that are nearly identical in appearance to those of Berken. Its hard seed proportion can reach 50%. Under favorable conditions or in the presence of powdery mildew pressure, however, it can yield up to 20% more than Berken.

2.7 Mungbean Breeding in Kenya

In Kenya, mungbean, with the average values of 290,050 ha area, 173,3694 ton production, and 684 kg/ha yield from 2016 to 2020, is a critical income security crop in marginal areas, which is the leading producer in Africa. But because the demand exceeds supply, Kenya has to be met with imports from Tanzania and Uganda. The mungbean production constraints in Kenya are diseases such as bacterial blight, bean anthracnose, and powdery mildew, as well as problems caused by insects like aphids, pod sucking bugs, and bruchids. So, the objectives of mungbean breeding are to develop improved marketable varieties with resistance to disease, improved grain yield, and desirable seed characteristics.

The mungbean improvement program was initiated at the Nationa Dryland Farming Research Station (NDFRS) in the late 1970s. By germplasm collection, introduction, and screening, two highly promising lines, KVr 22 and KVr 26, have been pre-released to farmers in the semiarid areas in 1983. In 1988, varieties KAT N26 and KAT N22 were released. But KAT N22 was rejected by farmers because of its yellow grain color. In 2012, the introduction and evaluation of advanced Worldveg lines for adaptability and yield potential were initiated. By multi-location test, participatory variety selection with farmers and national performance trial, three new varietes, Karembo, Biashara, and Ndengu tosha, were released for commercial production in 2017 with advantages such as >20% above best check, shiny and large green seed, early maturity with 70–75 days, and high Fe and Zn contents. In the future, promotion of the recently released varieties beyond the traditional mungbean production areas to accelerate the uptake is a priority.

2.8 Mungbean Breeding in the Republic of Korea

Mungbean has been grown in Korea for many centuries. It is the third most important legumes after soybean and adzuki bean. In 1986, mungbean was planted on 7815 ha, and the average yield was about 900ke/ha in the 1980s. Since 1981, the Korean government has allowed mungbean to be imported if the local harvest during the previous year was far below the local demand. Mungbean is imported primarily from Thailand and the Middle East. Korean sprout growers prefer small seed size (about 4.0 g/100 seeds with dull seed luster). It is estimated that Korea imports from 1000 t to 2000 t of mungbean annually. The breeding program for mungbean has been very small in scale, but the breeders in the Crop Experiment Station of the Rural Development Administration have been very effective and efficient. They have emphasized and well-organized adaptation testing of introduced breeding lines, mainly from the Asian Vegetable and Development Center (AVRDC). As a result, two AVRDC lines namely V3476 (a Philippine cultivar) and VC 1973 were released as Bangasa and Seonhwa-nogdu in 1980 and 1982, respectively. The two cultivars had the higher yield potential in many locations and well received by the farmers.

2.9 Mungbean Breeding in Pakistan

Mungbean is an important pulse crop in Pakistan, which accounts for about 35–40% of total pulse production in the country with about 30,0000 ha growing area and 870 kg/ha productivity. The cropping systems include wheat-mungbean-wheat, wheat-mungbean-cotton-wheat, wheat-mungbean-sesame-wheat, and intercropping with sugarcane and orchids. The major constraints of production in Pakistan are biotic stresses such as mungbean yellow mosaic virus disease, *Cercospora* leaf spot, white fly, hairy caterpillar, sucking insects, and bruchids.Therefore, its breeding aims include short duration, high yield, climate resilient, determinate growth habit with top pods in bunches, resistance to mungbean yellow mosaic virus, *Cercospora* leaf spot and bruchid, heat tolerance at flowering, reduced photoperiod sensitivity, synchronous maturity, long pods with bold, shiny seeds. By collaborating with the World Vegetable Center, 20 improved varieties with high yield, short duration (70–90 days), erect plant type, resistance to MYMV, bold seed size, and shiny green seed coat were released and extended in the country, including NM-92, NM94, NM-11, NM-16, NM-21, AZRI Mung-21, Jambo Mung, NCM-2013.

2.10 Mungbean Breeding in Sri Lanka

Mungbean can be considered the most popular grain legume in Sri Lanka. It is used either in split or boiled form. When split, a spiced curry is prepared and consumed with rice; when boiled, it is eaten with grated coconut for breakfast. Mungbean is also used in making a variety of sweets. In Sri Lanka, over 80% of the mungbean area is cultivated under rainfed conditions, resulting in the low average yield of 0.88 t/ha. Beanfly, pod borers, bruchids, collar rot, mungbean yellow mosaic virus (MYMV), *Cercospora* leaf spot, and powdery mildew are some of pests and diseases that cause substantial damage to the crop. The development of useful plant types with preferred characters which can fit into the farming system is closely related to increased yield potential and mungbean production level in Sri Lanka. So, the objectives of mungbean improvement include the following: to incorporate multiple disease resistance and field resistance to insect pests such as beanfly, pod borers, and bruchids; increase grain yield while maintaining or increasing percent protein, with preferred seed quality (large, green, glossy seeds); tailor varieties to fit into different moisture regimes; develop genotypes with increased ability to fix nitrogen under local conditions; and broaden the adaptation through better emergence, seedling vigor, root penetration, and tolerance to soil-moisture stress. Plant breeders in Sri Lanka have exploited the available genetic variability and developed many elite strains which have high seed yield, early and synchronous maturity, pest and disease tolerance, and rapid emergence with good seedling vigor. Among the varieties developed, MI-5 and T-77 have perfermed well throughout the country and variety (CES 55 X MI-3)-133F has showed early maturity (55–60 days) with tolerance to MYMV.

3 Challenges and Perspectives of Mungbean Breeding

Mungbean, a pulse crop, is commonly intercropped or relay-planted with other species due to its broad sowing window and nitrogen-fixing ability. Its short growth cycle and tolerance for poor soils also make it suitable for post-disaster recovery and reclamation of marginal land. Nutritionally, mungbean seeds are rich in protein, dietary fiber, vitamin B_2, calcium, and phosphorus, as well as functional compounds such as vitexin, isovitexin, and D-chiro-inositol, which are associated with heat-clearing, detoxifying, and hepatoprotective and vision-protective effects. Given these agronomic and nutritional attributes, mungbean plays a valuable role in improving soil structure, enriching human diets, and supporting agro-processing. It is therefore a crop with significant potential for further development worldwide.

The main constraints of mungbean production are determined by biotic stresses, abiotic stresses, production technology, and local policies. Among the biotic stresses, diseases are the most serious constraints, which affect the productivity and quality of seeds (Dikshit et al. 2020). Mungbean is prone to several viral, fungal, and bacterial diseases. Among the viral diseases, mungbean yellow mosaic disease (MYMD) is the most important, Which may cause 32–78% yields reduction in India (Kitsanachandee et al. 2013). Common fungal diseases are *Cercospora* leaf spot, powdery mildew, root rot, anthracnose, and blight. Bacterial diseases include halo blight, tan spot, *fusarium* wilt, and bacterial leaf spot. All bacterial pathogens are seed-borne and can persist in stubbles, so variety resistance is the key to the disease management. Insect pests, one of the major constraints in mungbean production, attack different parts of the plant including roots, shoots, flowers, and pods at different stage. The most important insects are bean/stem fly, white fly, aphids, pod borers, thrips, pod bugs, caterpillar, and bruchids. Abiotic stresses are those environmental factors that impact the plant biochemical and physiological processes to constrain productivity. The major abiotic stresses that effect the mungbean yield and quality include calcareous soil (iron deficiency chrolosis), saline alkali soil, water-logging, drought, and terminal heat (high and low). In some tropical place, mungbean may also suffer from prolonged rainy season resulting in excess soil and atmospheric moisture leading to pre-harvest sprouting. Pod-shattering is another constraint which affects mungbean production, especially the varieties that have prolonged flowering.

The constraints of marketing for mungbean are quality of the seeds, including commercial quality (appearance, such as shiny or dull, seed size), nutritional quality (nutrition contents such as protein, folic acid, vitexin, isovitexin, zinc, iron), and processing quality (sprout quality such as sprout yield, length, thickness, whiteness, and crispness). Compared with soybean, mungbean has a widely acceptable sensory profile and excellent nutritional properties with high level of protein and amino acids. The DIAAS (Digestible Indispensable Amino Acid Score) is 93 that means a good quality of protein. It rarely has allergy incidence with strong regulatory precedent and can enhance muscle recovery and bioavailability.

The objectives of mungbean breeding should tackle the above constraints of biotic, abiotic, and production requirements and marketing. So, the major breeding objectives of mungbean should include the following: (1) high and stable yield (with 2.0–2.5 t/ha of optimal yield), (2) ideal plant types for cropping system and mechanized harvest: short maturity, medium plant height (60-80 cm), determinate growth habit and synchronous maturity, more pods at top of plant and non-shattering habit, (3) resistance to multiple disease: including mungbean yellow mosaic disease (MYMD), leaf spot, powdery mildew, root rot, anthracnose, blight, tan spot, and fusarium wilt, (4) tolerance to abiotic stresses, including calcareous soil (iron deficiency chrolosis), saline alkali soil, water-logging, drought, and terminal heat (high and low), (5) good quality of seeds: depends on the market demands, big seed or small seed, and shiny or dull, and (6) good quality for food processing and nutrient fortifications, such as good-quality bean sprout and high contents of protein, zinc, iron, folate, vitexin, etc.

Suitable and efficient breeding methods relying on the conditions are crucial for achieving breeding objectives. The common methods that have been deployed are introduction, pedigree selection, crossbreeding between varieties, and physical and chemical mutation. Introduction and pedigree selection from indigenous and exotic germplasm or elite lines as well as landraces have layed an important role in the development of superior cultivars of mungbean. VC 1973A, an elite line developed by AVRDC, is a very good example, which was released and popularized in many countries in the world. Crossbreeding includes intraspecific cross and interspecific cross. For the mungbean breeding, the most successful method now is the intraspecific cross, which is simple, effective, and easily adopted. Up to now, 141 new varieties have been developed in China by intraspecific cross, leading to 12–38.6% increase of yield and solving problems such as long growth period, creeping, lodging, poor adaptability, and pod shattering. Among them, Zhonglv 6, Zhonglv 7, Jilv 15, Jilv 17, Su Lv 5, Su Lv 6, and Jin Lv 7 are resistant to bruchid (Wu et al. 2022), avoiding bruchid damage during storage. Jilv 19, Jilv 10, and Zhonglv 12 have compact plant types, small branching angles, high podding positions, determinate growth habit, and synchronous maturity, making them suitable for mechanized harvesting(Tian et al. 2021). So, intraspecific cross will still be the main means of mungbean breeding in the future. Interspecific hybridization can import some excellent traits from other *vigna* species, such as MYMV resistance from black gram (*Vigna mungo*) and bruchid resistance from rice bean (*Vigna umbellata*). Despite numerous attempts of interspecific hybridization between mungbean and relatives species, the actual release of new cultivars from distant crosses has remained limited. Mutation breeding refers to the induction of genetic variations in plants through physical and chemical factors, selection of individual plants from the mutant population, and then development of new varieties. Valuable mutants can be obtained by mutation breeding in a relatively short period of time, significantly improving the efficiency of plant breeding. Mutation through physical (e.g., ^{60}Co-γ) and chemical (e.g., EMS) methods has been successfully used in mungbean breeding. However, mutation breeding has some drawbacks, including the lack of directionality of mutations and the problem of fewer favorable mutations. It is difficult to simultaneously improve multiple traits. So, mutation breeding is only an auxiliary means for future mungbean breeding.

Conventional breeding methods such as crossbreeding present several limitations, including long development cycles, low efficiency, and challenges in effectively combining multiple desirable traits. In contrast, molecular marker-assisted selection (MAS) offers significant advantages by shortening the breeding timeline, accelerating variety development, and enhancing overall breeding precision, and efficiency. Now, a few of molecular markers are available for resistant selection of some diseases and pests, such as MYMV, powdery mildew, CLS, and bruchids (Somta et al. 2022; Alam et al. 2014; Kaewwongwal et al. 2017; Yundaeng et al. 2021).

With the continuous identification of molecular markers associated with critical agronomic traits like disease resistance and yield, MAS is poised to play a pivotal role in advancing mungbean breeding. Meanwhile, the routine application of more complex technologies, including genomic selection and gene editing, is anticipated to require further development, as robust, crop-specific technical frameworks for mungbean have yet to be fully established.

4 Conclusion

Mungbean, as a pulse, has valuable roles in soil structural improvement, human dietary enrichment, and the processing and utilization of agricultural products. It has the trend toward a gradual increase in planting area globally. The new varieties play an important role in the global mungbean production. The objectives of mungbean breeding should tackle the constraints of biotic, abiotic, production requirements and marketing, including high and stable yield, ideal plant types for cropping system and mechnized harvest, resistance to multiple disease, tolerance to abiotic stresses, good market quality, good quality for food processing and nutrient fortifications. Suitable and efficient breeding methods relying on the above conditions are crucial for achieving breeding objectives. The most successful method of mungbean breeding now is the intra-specific cross, and will still be the main means of mungbean breeding in the future. With the development of available molecular markers, molecular marker-assisted selection (MAS) would be a powerful strategy for mungbean breeding in the near future. Other molecular technologies such as whole genome selection and gene editing will still take time to be applied, because of lack of well-established technical systems in mungbean.

References

Alam AKMM, Somta P, Srinives P (2014) Identification and confirmation of QTLs controlling resistance to mungbean yellow mosaic disease in mungbean (*Vigna radiata* (L.) Wilczek). Mol Breed 34:1497–1506

Ali M, Malik IA, Sabir HM, Ahmad B (1997) The mungbean green revolution in Pakistan. Technical Bulletin No. 24, AVRDC. Shanhua, Taiwan, p 66

AMA (2014) Australian mungbean industry strategic plan 2015-2019. Queensland, Australian Mungbean Association

Anonymous (2021). Greengram Outlook Report. Available online at: https://angrau.ac.in/downloads/AMIC/GREENGRAM%20OUTLOOK%20REPORT %20-January%20to%20May%20 2021.pdf. Accessed 7 Apr 2022

Ballon FB, Legaspi BM, Catipon EM (1977) Mungbean varietal improvement program of the bureau of plant industry. In: Cowell R (ed) The first international mungbean symposium. AVRDC, Shanhua, pp 217–219

Bose RD (1932) Studies in Indian pulses. IV Mung or Green gram (*Phaseolus aureus* Roxb). Indian J Agric Sci 2:607–624

Cao D, Li H, Yi J, Zhang J, Chen H, Cao J et al (2011) Antioxidant properties of the mungbean flavonoids on alleviating heat stress. PLoS One 6:e21071

Catipon EM, Legaspi BM, Jarilla FA (1988) Development of mungbean varieties from AVRDC linesforthe Philippines. In: Shanmugasundaram S, McLean BT (eds) Mungbean: proceeding of the second international symposium. AVRDC, Shanhua, pp 88–97

Chankaew A, Somta P, Isemura T, Tomooka N, Kaga A, Vaughan DA, Srinives P (2013) QTL mapping reveals conservation of major and minor loci for powdery mildew resistance in four sources of resistance in mungbean [*Vigna radiata* (L.) Wilczek]. Mol Breed 32:121–130

Dikshit HK, Mishra GP, Somta P, Shwe T et al (2020) Classical genetics and traditional breeding in mungbean. In: Nair RM, Schafleitner R, Lee SH (eds) The mungbean genome. Springer International Publishing, Cham, pp 43–54

Domingo A (2023) Mung bean production as a source of livelihood among farmers in a municipality in The Philippines: challenges and opportunities. Asian J Agric Rural Dev 13(2):130–137

Fernandez GCJ, Shanmugasundaram S (1988) The AVRDC mungbean improvement programme: the past, present and future. In: Proceedings of the 2nd International Mungbean Symposium. Asian Vegetable Research and Development Center, Tainan, pp 58–70

Kaewwongwal A, Chen J, Somta P, Kongjaimun A, Yimram T, Chen X, Srinives P (2017) Novel alleles of two tightly linked genes encoding polygalacturonase-inhibiting proteins (VrPGIP1 and VrPGIP2) associated with the Br locus that confer bruchid (Callosobruchus spp.) resistance to mungbean (*Vigna radiata*) accession V2709. Front Plant Sci. https://doi.org/10.3389/fpls.2017.01692

Kitsanachandee R, Somta P, Chatchawankanphanich O, Akhtar KP, Shah TM, Nair RM, Bains TS, Sirai S, Kaur L, Srinives P (2013) Detection of quantitative trait loci for mungbean yellow mosaic India virus (MYMIV) resistance in mungbean (*Vigna radiata* (L.) Wilczek) in India and Pakistan. Breed Sci 63:367–373

Laosatit K, Yimram T, Keawwongwal A, Srichan M, Amkul K, Tanadul O, Masmeatathip R, Somta P (2024) Development of pyramided mung bean lines carrying resistance genes for Cercospora leaf spot disease and bruchids. Chilean J Agric Res 84:644–652

MAOLI (2019) Myanmar agriculture in brief, 2018 and 2019. The Ministry of Agriculture, Livestock and Irrigation (MAOLI), Naypyidaw

Ministry of Agriculture. Livestock and Irrigation (2016) Myanmar agriculture at a glance 2016. Department of Planning, Ministry of Agriculture, Livestock and Irrigation. Nay Pyi Taw

Nair R, Schreinemachers P (2020) Global status and economic importance of mungbean. In: Nair RM, Schafleitner R, Lee SH (eds) The mungbean genome. Springer International Publishing, Cham, pp 1–8

Schreinemachers P, Sequeros T, Rani S, Rashid MA, Gowdru NV, Rahman MS, Ahmed MR, Nair RM (2019) Counting the beans: quantifying the adoption of improved mungbean varieties in South Asia and Myanmar. Food Secur 11(3):623–634

Shanmugasundaram S, Keatinge JDH, d'Arros Hughes J (2009) The mungbean transformation. Diversifying crops, defeating malnutrition. In: Spielman DJ, Pandya-Lorch R (eds) Millions fed: proven successes in agriculture development. International Food Policy Research Institute, Washington, DC, pp 103–108

Somta P, Laosatit K, Yuan X, Chen X (2022) Thirty years of mungbean genome research: where do we stand and what have we learned? Front Plant Sci 13:944721

Somta P, Sorajjapinun W, Yimram T, Tanadul O, Laosatit K, Srinives P (2024) Registration of 'KUML4' and 'KUML8' mungbean cultivars with high yield and large seeds. J Plant Regis 18:33–41

Tang D, Dong Y, Ren H, Li L, He C (2014) A review of phytochemistry, metabolite changes, and medicinal uses of the common food mung bean and its sprouts (*Vigna radiata*). Chem Cent J 8:4. https://doi.org/10.1186/1752-153X-8-4

Tian J, Cheng X, Fan B, Wang L, Liu J et al (2021) Current situation and development trend of mungbean varieties in China. Crop J 37(6):15–21

Wu R, Zhang Q, Lin Y, Chen J, Somta P, Yan Q, Xue C, Liu J, Chen X, Yuan X (2022) Marker-assisted backcross breeding for improving bruchid (Callosobruchus spp.) resistance in mung bean (*Vigna radiata* L.). Agronomy 12(6):1271

Yimram T, Somta P, Srinives P (2009) Genetic variation in cultivated mungbean germplasm and its implication in breeding for high yield. Field Crop Res 112:260–266

Yundaeng C, Somta P, Chen J, Yuan X, Chankaew S, Chen X (2021) Fine mapping of QTL conferring Cercospora leaf spot disease resistance in mungbean revealed TAF5 as candidate gene for the resistance. Theor Appl Genet 134:701–714

Wild Species of *Vigna* Crosses Utilized in Mungbean Improvement

Muthiyan Pandiyan, Sivakumar Paramasivam, A. Gopikrishanan, A. Mahalingam, A. Krishanveni, and K. R. Jahanmohan

Abstract

Interspecific cross is a significant method for transferring desirable genes from wild to cultivated *Vigna* species. A varying level of crossability has been reported in *Vigna radiata* and other wild *Vigna* species. A total of 175 flowers were emasculated, with the highest number recorded in *Vigna radiata* var. *sublobata* and the lowest number (78) in *V. pilosa* crosses. A total of 141 flowers were pollinated in the *V. radiata* × *V. radiata* var. *sublobata* cross, while 72 flowers were pollinated in *V. pilosa* cross combinations. Thirty-five crossed pods were obtained in *V. radiata* var. *sublobata*, whereas only six pods were recorded in *V. capensis* and *V. khandalensis*. The highest number of seeds (240) was produced in *V. radiata* var. *sublobata*, while the lowest number (25) was observed in *V. pilosa*. Seed germination was highest in *V. radiata* var. *sublobata*, with 190 seeds germinated, and lowest in *V. bournia*, with only five seeds. The maximum crossed-seed germination percentage (79.2%) was recorded in *V. radiata* × *V. radiata* var. *sublobata*, whereas the minimum (12.5%) was noted in *V. bournia*. Plant survival was

M. Pandiyan (✉) · S. Paramasivam · K. R. Jahanmohan
Dr. M. S. Swaminathan Agricultural College and Research Institute, Tamil Nadu Agricultural University, Eachangkottai, Thanjavur, Tamil Nadu, India

A. Gopikrishanan
Agricultural Research Station, Tamil Nadu Agricultural University, Virinjipuram, Vellore, Tamil Nadu, India

A. Mahalingam
Regional Research Station, Tamil Nadu Agricultural University, Vridhachalam, Cuddalore, Tamil Nadu, India

A. Krishanveni
Agricultural Research Station, Tamil Nadu Agricultural University, Vazhavachanur, Tiruvannamalai, Tamil Nadu, India

S. Paramasivam et al. (eds.), *Mungbean*,
https://doi.org/10.1007/978-981-95-7637-1_3

highest in *V. radiata* var. *sublobata* (170 plants) and lowest in *V. khandalensis* (three plants). At harvest, 152 plants were productive in *V. radiata* var. *sublobata*, while productivity was lowest in *V. khandalensis*. Hybrid breakdown was highest (89.4%) in the *V. radiata* × *V. radiata* var. *sublobata* cross and lowest (33.3%) in *V. radiata* × *V. khandalensis*. Hybrid sterility was highest (66.7%) in *V. radiata* × *V. khandalensis* and lowest (10.6%) in the *V. radiata* × *V. radiata* var. *sublobata* cross. The percentage of lethality among interspecific hybrids varied from 10.6 in *V. radiate* var *sublobata* to 66.7. The progressive results of this study provide an opportunity for the successful development of interspecific hybrids in *Vigna* breeding program.

Keywords

Interspecific hybridization · Crossability · Hybrid breakdown · Hybrid lethality · Hybrid sterility · *Vigna radiate*

1 Introduction

Asian *Vigna* species show rich diversity in India and other parts of Asia. The genus *Vigna* is divided into six subgenera, Sigmoidotropis, *Haydonia*, *Sigmoidotropis Plectotropis, Vigna*, *Ceratotropis*, and *Lasiopron*. The subgenus *Sigmoidotropis* within the genus *Vigna* includes about 90 species and is divided into six subgenera, as reported by Baudoin et al. (1989). The New World *Vigna* subgenus *Sigmoidotropis* is more closely related to *Phaseolus* (Delgado-Salinas et al. 2022). The Catiang group is related to cowpea, while the Reticulatae group is closely related to the subgenus *Plectrotropis*. The Asian subgenus *Ceratotropis* forms a distinct group. The subgenus *Haydonia*, found in Africa, may represent the most primitive type within *Vigna*. The cultivated *Vigna* species include *Vigna aconitifolia*, *Vigna angularis* (adzuki bean), blackgram (*Vigna mungo*), *Vigna radiata* (mungbean), *Vigna umbellata* (rice bean), *Vigna trilobata* (phillipsara), and *Vigna unguiculata* (cowpea, grain type). The vegetable types of cowpea include *V. unguiculata* subsp. *sesquipedalis* and *V. unguiculata* subsp. *cylindrica*.

The following wild *Vigna* species were cultivated in the tribal areas for their consumption.

1. Greengram wild progenitor *V. radiata* var. *sublobata*
2. Blackgram wild progenitor *V. mungo* var. *silvestris*
3. Adzukibean wild progenitor *V. angularis* var. *niponensis*
4. Ricebean wild *V. umbellata* var. *gracilis*
5. Cowpea wild progenitor *Vigna ungiculata* subsp. *dekindtiana, Vigna ungiculata* subsp. *mensensis*, *Vigna ungiculata* subsp. *vexillata*

2 Origin of *Vigna* species

The origin of species generally occurs in two ways, viz., from wild progenitors or through evolutionary processes such as mutation, hybridization, migration, gene flow, and genetic drift, as evidenced by archaeological records (Pratap and Kumar 2011). Mungbean and urdbean were earlier considered to have originated in India (Vavilov 1926; Zukovskij 1962). A wide diversity of wild and cultivated forms of mungbean has been reported (Baudoin and Maréchal 1988). In urdbean, both weedy and cultivated forms occur in the Western Ghats, indicating long-term domestication in this region (Singh et al. 1974; Chandel et al. 1984; Lawn and Cottell 1988). Cowpea is believed to have originated in West and Central Africa, where several weedy forms occur in forest and savannah regions (Harlan and de Wet 1971; Rawal 1975). Archaeological records suggest that cowpea was under cultivation as early as 3000 BC. Ricebean has been traced to Asia (Vavilov 1926), and *Vigna minima* is regarded as the wild progenitor of ricebean (Gopinathan and Babu 1986). *Vigna minima (Roxb) Ohwi* and *Ohashi* has also been referred to as *Vigna delzelliana* var. *gracilis* (Marechal et al. 1978). Moth bean, Vigna aconitifolia, was documented by Smartt (1985) and Vavilov (1926). Maréchal et al. (1978) identified Sri Lanka and Pakistan as centers of diversity for *Vigna* crops. Adzuki bean is considered to have originated in the Chinese center (Vavilov 1926). The wild form of adzuki bean, *Vigna angularis* var. *nipponensis*, has been reported by Yamaguchi (1992) and Kaga et al. (2008). Weedy forms of adzuki bean have also been described (Vaughan et al. 2005).

3 Different Gene Pools of Mungbean and Urdbean

Primary gene pools, secondary gene pools, and tertiary gene pools of some cultivated and their wild species are as follows:

Mungbean

Primary gene pool is *Vigna radiata* var. *sublobata*

Secondary gene pool is *Vigna mungo* var. *silvestris*, *Vigna trilobata*, and *Vigna aconotifolia*

Tertiary gene pool is *Vigna angularis, Vigna dalzelliana, Vigna glabrescens, Vigna grandis, Vigna umbellate*, and *Vigna vexillata*

(Chandel and Laster 1991; Dana and Karmakar 1990; Smartt 1985; Kumar et al. 2004)

Urdbean

Primary gene pool is *V. mungo* var. *mungo*, *V. mungo* var. *silvestris*

Secondary gene pool is *Vigna radiata* var. *sublobata*, *Vigna settulosa, Vigna trilobata*, and *Vigna aconotifolia*

Tertiary gene pool is *Vigna angularis, Vigna dalzelliana, Vigna glabrescens, Vigna grandis, Vigna umbellate,* and *Vigna vexillata*

(Chandel and Laster 1991; Dana and Karmakar 1990; Kumar et al. 2004)

4 Status of *Vigna* Species in Global and India Level

India is a significant center of diversity for *Vigna* species, which are distributed across diverse terrains, from coastal regions to high-altitude Himalayan areas. The number of *Vigna* species available at the global level is listed below.

Vigna abyssinica, Vigna aconitifolia, Vigna acuminata, Vigna adenantha, Vigna afzelii, Vigna alba, Vigna ambacensis, Vigna andongensis, Vigna angivensis, Vigna angularis, Vigna angustifolia, Vigna angustifoliolata, Vigna anomala, Vigna antillana, Vigna antunesii, Vigna appendiculata, Vigna aristata, Vigna asparaginea, Vigna baoulensis, Vigna benthami, Vigna benuensis, Vigna bequaertii, Vigna bosseri, Vigna bourneae, Vigna brachycalyx, Vigna brachycarpa, Vigna brachystachys, Vigna brasilica, Vigna brasiliensis, Vigna briartii, Vigna buchneri, Vigna bukobensis, Vigna burchellii, Vigna caesia, Vigna calcarata, Vigna campestris, Vigna candida, Vigna canescens, Vigna capensis, Vigna capitata, Vigna caracalla, Vigna carinalis, Vigna catiang, Vigna catjang, Vigna chiovendae, Vigna clarkei, Vigna coerulea, Vigna comosa, Vigna congoensis, Vigna crinita, Vigna cylindrica, Vigna dalzelliana, Vigna dauciformis, Vigna davyi, Vigna debanensis, Vigna debilis, Vigna decipiens, Vigna dekindtiana, Vigna desmodioides, Vigna diffusa, Vigna dimorphophylla, Vigna dinteri, Vigna dolichoides, Vigna dolichonema, Vigna dolmitica, Vigna elegans, Vigna esculenta, Vigna exillata, Vigna filicaulis, Vigna firmula, Vigna fischeri, Vigna fragrans, Vigna friesiorum, Vigna frutescens, Vigna galpinii, Vigna.gangetica, Vigna gazensis, Vigna gentryi, Vigna glabra, Vigna glabrescens, Vigna glandulosa, Vigna golungensis, Vigna gracilicaulis, Vigna gracilis, Vigna grahamiana, Vigna grandis, Vigna halophila, Vigna hapalantha, Vigna harmsiana, Vigna harmsii, Vigna hastat, Vigna.hastifolia, Vigna haumaniana, Vigna helicopus, Vigna heterophylla, Vigna hirsuta, Vigna hirta, Vigna hirtella, Vigna hispida, Vigna holosericea, Vigna holstii, Vigna homblei, Vigna.hookeri, Vigna.hosei, Vigna huillensis, Vigna hundtii, Vigna hygrophila, Vigna incana, Vigna inga, Vigna jaegeri, Vigna juncea, Vigna junodii, Vigna juruana, Vigna kassneri, Vigna katangensis, Vigna keniensis, Vigna keraudrenii, Vigna khandalensis, Vigna kirkii, Vigna kokii, Vigna kotschyi, Vigna lanceolata, Vigna lancifolia, Vigna lasiocarpa, Vigna latidenticulata, Vigna laurentii, Vigna lebrunii, Vigna ledermannii, Vigna leptodon, Vigna linearifolia, Vigna linearis, Vigna lobata, Vigna lobatifolia, Vigna lobelifolia, Vigna lonchophylla, Vigna longepedunculata, Vigna longifolia, Vigna longiloba, Vigna longissima, Vigna lozanii, Vigna lucens, Vigna lutea, Vigna luteola, Vigna macrantha, Vigna macrodon, Vigna macrorrhyncha, Vigna malayana, Vigna malosana, Vigna maranguensis, Vigna marchali, Vigna marginata, Vigna marina, Vigna matengoana, Vigna megatyla, Vigna membranacea, Vigna membranaceoides, Vigna mendesii, Vigna mensensis, Vigna micrantha, Vigna microsperma, Vigna mildbraedii, Vigna minima, Vigna monachalis, Vigna monophylla, Vigna multiflora, Vigna multinervis, X, Vigna myrtifolia, Vigna nakashimae, Vigna nervosa, Vigna neumannii, Vigna nigerica, Vigna nigritia, Vigna nilotica, Vigna ntemensis, Vigna nuda, Vigna owahuensis, Vigna oahuensis, Vigna oblonga, Vigna oblongifolia, Vigna occidentalis, Vigna oligosperma, Vigna opisotricha, Vigna ornata, Vigna oryzina,

Vigna ostinii, Vigna oubanguensis, Vigna paludosa, Vigna papuana, Vigna paraguariensis, Vigna parkeri, Vigna parviflora, Vigna parvifolia, Vigna peduncularis, Vigna perrieriana, Vigna phaseoloides, Vigna phoenix, Vigna pilosa, Vigna platyloba, Vigna polytricha, Vigna polytyla, Vigna pongolensis, Vigna populnea, Vigna praecox, Vigna proboscidella, Vigna procera, Vigna pseudolablab, Vigna pseudotriloba, Vigna pubescens, Vigna pubigera, Vigna punctata, Vigna pusilla, Vigna pygmaea, Vigna racemosa, Vigna radiata, Vigna radicans, Vigna ramanniana, Vigna reflexopilosa, Vigna repens, Vigna reticulata, Vigna retusa, Vigna rhomboidea, Vigna richardsiae, Vigna ringoetii, Vigna riukiuensis, Vigna rotundifolia, Vigna sahyadriana, Vigna sandwicensis, Vigna scabra, Vigna scabrida, Vigna schimperi, Vigna schlechteri, Vigna schliebenii, Vigna schottii, Vigna schweinfurthii, Vigna senegalensis, Vigna sepiaria, Vigna serotina, Vigna sesquipedalis, Vigna silvestris, Vigna sinensis, Vigna somaliensis, Vigna spartioides, Vigna speciosa, Vigna spectabilis, Vigna stenocarpa, Vigna stenodactyla, Vigna stenophylla, Vigna stipulacea, Vigna stipulata, Vigna strobilophora, Vigna strophiolata, Vigna stuhlmannii, Vigna suberecta, Vigna subhastata, Vigna sublobata, Vigna subterranea, Vigna sudanica, Vigna sulphurea, Vigna syndwicensis, Vigna taubertii, Vigna tenuis, Vigna thonningii, Vigna tisseranti, Vigna tisserantiana, Vigna tisserantii, Vigna tonkinensis, Vigna tranquebarica, Vigna trichocarpa, Vigna triloba, Vigna trilobata, Vigna triphylla, Vigna truxillensis, Vigna tuberosa, Vigna uliginosa, Vigna ulugurensis, Vigna umbellata, Vigna unguiculata, Vigna valetoni, Vigna variegata, Vigna venulosa, Vigna venusta, Vigna verdcourtii, Vigna verticillata, Vigna vexillata, Vigna vignoides, Vigna villosa, Vigna violacea, Vigna virescens, Vigna wightii, Vigna wilmsii, Vigna wittei.

5 Diversity, Distribution of Wild and Cultivated Species

India is a country exhibiting considerable diversity and serving as a center of origin for *Vigna* species, which are abundantly found in the Western Ghats, north-western Himalayas, and northeastern region.

Several wild *Vigna* species such as *V. radiata* var. *sublobata, V. mungo* var. *silvestris, V. umbellata,* and *Vigna hainiana* have wide distributions, while others like *Vigna settulosa* and *Vigna khandalensis* are more localized. Commonly cultivated *Vigna* species include *Vigna unguiculata* subsp. *sesquipedalis* (yard long bean) and *Vigna unguiculata* (cowpea) (Fig. 1).

6 Importance of Crop Improvement

Wild *Vigna* species possess valuable traits such as pest resistance, disease resistance, and drought tolerance. These traits make them important resources for crop improvement programs.

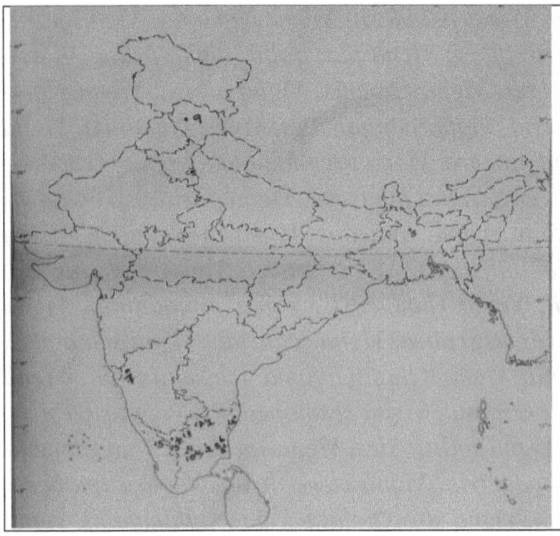

Species	Place of collection
1. V.radiata var. sublobata	Vellore and Perambalur
2. V. mungo var. silvestris	Perambalur, Goa
3. V. umbellata	Manipur
4. V. hainiana	Palakad, Thrissure
5. V. trilobata	Chidambaram, Cuddalore
6. V. acconitifolia-	Chidambaram, Tindivanam
7. V. settulosa	Goa, manipur
8. V. stipulacea	Chidambaram
9. V. angularis	Himachala Pradesh
10. V. Vexillata	Kodaikanal
11. V. glabrascense	Kodaikanal
12. V. marina	Chennai
13. V. bournia	Palakad
16.V. capensis	Palani hills
15. V. pilosa	Palakad
16. V. khandalensis	Trichur
17. Vigna radiata	Vellore

Fig. 1 Map showing the locations from where *Vigna* species were collected in India

6.1 Conservation Concerns

Some wild *Vigna* species, particularly those found in specific regions like the Savalghat hill, require immediate conservation efforts to prevent extinction.

6.2 Research Focus

Research on *Vigna* species in India includes studies on their diversity, genetic resources, morphological characteristics, and potential for crop improvement.

6.2.1 Interspecific Hybridization in *Vigna* Species

Interspecific hybridization plays an important role in producing new gene combinations associated with plant characteristics. Interspecific hybridization plays an important role in producing new genes associated with plant characteristics (Asthana et al. 1994). The introgression of wild and related species of *Vigna* has also been reported from the 1970s (Ahuja and Singh 1977; Ahn and Hartmann 1978a, b). Some germplasm accessions include the cultivated forms of *Vigna* species, which have desirable plant traits and are resistant to pests and diseases. (Umamaheshwari 2002; Somta et al. 2007). Wild *Vigna* species, namely *V. trilobata*, *V. vexillata*, *V. silvestris*, *V. hainiana*, and *V. umbellata*, have contributed to the improvement of *Vigna radiata* and *Vigna mungo*. In a similar context, apart from the cultivated species of *Vigna*, wild species such as *Vigna trilobata*, *Vigna grandis*, *Vigna dalzelliana*, *V. vexillata*, *Vigna radiata* var. *sublobata*, and *Vigna mungo* var. *silvestris* (Fig. 2) are widely used in developing high-yielding genotypes with resistance to biotic and abiotic stresses. Although crossability

Fig. 2 Photographs of flowering and podding stage of various *Vigna* species (a) *Vigna radiata sublobata,* (**b**) *Vigna settulosa,* (**c**) *Vigna hainiana,* (**d**) *Vigna khandalensis,* (**e**) *Vigna mungo silvestris,* (**f**) *Vigna acconitifolia,* (**g**) *Vigna trilobata,* (**h**) *Vigna bournea,* (**i**) *Vigna umbellata,* and (**j**) *Vigna vexillata*

between these species has been achieved, reproductive barriers still pose challenges to breeders (Dana and Karmarkar 1990; Siriwardhane et al. 1991; Pandiyan et al. 2010a; 2012a, b; Bhanu et al. 2017; Chitra et al. 2018; Lekhi et al. 2018).

6.2.1.1 Materials and Methods

In this study, 17 *Vigna* species were used, namely *Vigna radiata* var. *sublobata*, *Vigna mungo* var. *silvestris*, *Vigna setulosa*, *Vigna hainiana*, *Vigna minima*, *Vigna trilobata*, *Vigna aconitifolia*, *Vigna umbellata*, *Vigna angularis*, *Vigna vexillata*, *Vigna glabrescens*, *Vigna marina*, *Vigna khandalensis*, *Vigna pilosa*, *Vigna capensis*, *Vigna radiata*, and *Vigna mungo*. These species were used to develop direct crosses with *V. radiata* as the female parent. The experiment was conducted at the Agricultural Research Station, Virinjipuram, TNAU, Vellore, from 2014 to 2017. Hybridization techniques were conducted according to Boling et al.'s (1961) method. Emasculation was performed on the evening preceding the experiment between 4:00 and 6:00 PM. Pollination was performed the next morning between 6:00 and 8:00 AM. A large number of flowers were emasculated and pollinated.

Hybrid lethality and hybrid sterility have also been recorded in interspecific hybrids between various wild *Vigna* species and *V. radiata*.

$$\text{Hybrid lethality} = \frac{\text{No.of plants died}}{\text{No.of seeds germinated}} \times 100$$

$$\text{Hybrid breakdown} = \frac{\text{Hybrid in viability}}{\text{No.of plants germinated}} \times 100$$

$$\text{Hybrid sterility} = \frac{\text{No.of plants productive at harvest}}{\text{No.of plants survived}} \times 100$$

6.2.1.2 Results

Vigna species are rich in protein and are important pulse crops. More than 200 *Vigna* species are known. In this study, 17 species were used for crossing to create variability in yield traits and in biotic and abiotic stress resistance (Table 1).

Interspecific crosses within the same genus have made significant contributions to crop improvement by enabling the transfer of genes from wild species into cultivated species. Among the *Vigna* species, *Vigna radiata* has the ability to cross with both wild and cultivated *Vigna* species, often serving as the female parent.

In F_1 hybrids, crossability barriers such as hybrid inviability, hybrid sterility, and hybrid breakdown occur, which pose challenges to breeders' efforts in achieving successful hybridization between *Vigna* species. Wide hybridization among *Vigna* species contributes to the broadening of the genetic base and improvement of traits such as resistance to pod shattering, dormancy breakdown, and tolerance to biotic and abiotic stresses in cultivated *Vigna* species. In this study, 17 *Vigna* species were used, including *Vigna radiata* and 16 wild *Vigna* species, with greengram (*V. radiata*) as the female parent and the wild species as male parents.

The results of direct crosses, including crossability percentage, hybrid sterility, and hybrid breakdown of *V. radiata* crossed with wild *Vigna* species, are presented in Table 2. A total of 175 flowers were emasculated in *Vigna radiata*, the highest among the species, while the lowest, 78 flowers, were emasculated in *Vigna pilosa*. A total of 141 flowers were pollinated in the cross *Vigna radiata* × *Vigna radiata*

Table 1 Wild *Vigna* species used for interspecific hybridization

Sl. No	Species	Characters to be transferred	Place of collection
1.	*V. radiata* var. *sublobata*	More no. of pods, earliness, small seed size, brown, black ash color seeds	Vellore and Perambalur
2.	*V. mungo* var. *silvestris*	Terminal pod, less foliage, earliness	Perambalur, Goa
3.	*V. umbellata*	Complete resistance to MYMV and bruchids, drought, and moisture tolerance	Manipur
4.	*V. hainiana*	Earliness, more number of pods, prostrate nature	Palakad, Thrissure
5.	*V. trilobata*	Lengthy peduncle, more number of pods, versatile to all environment, small size black seed	Chidambaram, Cuddalore
6.	*V. acconitifolia-*	More number of pods, drought tolerance, pod borer resistance, short day plant	Chidambaram, Tindivanam
7.	*V. settulosa*	Earliness, more number of pods/cluster, tendrilness, cluster bearing	Goa, Manipur
8.	*V. stipulacea*	Earliness, short plant type, resistance to waterlogging, bushy growth	Chidambaram
9.	*V. angularis*	Erect type, more number of pods, resistance to high moisture stress, different seed colors	Himachal Pradesh
10.	*V. Vexillata*	Tuber, bruchids resistance, lengthy pods, more number of seeds/pod hairy pods	Kodaikanal
11.	*V. glabrascense*	Tetraploid, long duration, drought tolerance, more number of pods, glabrous leaves	Kodaikanal
12.	*V. marina*	Drought tolerance, leafy, long duration, salt tolerant	Chennai
13.	*V. bournia*	Climbing type, more number of pods, drought tolerance, creeper type	Palakad
14.	*V. capensis*	More number of pods, resistance to high moisture, violet color flower	Palani hills
15.	*V. pilosa*	Climbing type, fruit borer tolerance, velvet color pods	Palakad
16.	*V. khandalensis*	Earliness, synchronized maturity, sturdy plant type	Trichur
17.	*Vigna radiata*	Widely adaptable high-yielding variety, good quality for human consumption (high protein, iron, and digestibility)	Vellore

var. *sublobata*, while 72 flowers were pollinated in the *Vigna pilosa* cross combination. Thirty-five percent of crossed pods were harvested in *Vigna radiata* var. *sublobata*, while the lowest, 6%, was observed in *V. capensis* and *V. khandalensis*.

The highest number of seeds (240) was obtained in *Vigna radiata* var. *sublobata*, whereas the lowest number (25) was recorded in *V. pilosa*. Nearly 190 seeds germinated in *Vigna radiata* var. *sublobata* and 5.0 of seeds germinated in *V. bournia*.

Table 2 Biomterical, hybrid lethality, hybrid breakdown, and hybrid sterility of different wild *Vigna* species

Name of the species	No. of flower emasculated	No. of flowers crossed	Pods Set	Crossability percent age (%)	No. of seeds obtained	Seeds germinated	Germination (%)	Plants survive d	Productive plant at harvest	Hybrid lethality	Hybrid break down	Hybrid sterility (%)
V. radiata var. *sublobata*	175	141	35	31.9	240	190	79.2	170	152	10.6	89.4	10.6
V. mungo var. *silvestris*	150	105	27	25.7	175	130	74.3	101	86	14.9	85.1	14.9
V. umbellata	137	100	18	18.0	100	32	32.0	18	10	44.4	55.6	44.4
V. hainiana	127	85	15	17.6	74	38	51.4	26	18	30.8	69.2	30.8
V. trilobata-	133	84	20	23.8	85	37	43.5	27	16	40.7	59.3	40.7
V. acconitifolia	135	102	20	19.6	110	55	50.0	35	24	31.4	68.6	31.4
V. settulosa	110	80	17	21.3	100	70	70.0	54	42	22.2	77.8	22.2
V. stipulacea	130	98	20	20.4	98	48	49.0	35	22	37.1	62.9	37.1
V. angularis	102	85	12	14.1	62	25	40.3	18	12	33.3	66.7	33.3
V. Vexillata	127	95	10	10.5	54	9	16.7	5	3	40.0	60.0	40
V. glabrascense	150	119	17	14.3	95	35	36.8	25	19	24.0	76.0	24
V. marina	113	102	10	9.8	93	29	31.2	16	9	43.7	56.3	43.8
V. bournia	128	105	11	10.5	40	5	12.5	5	2	60.0	40.0	60
V. capensis	120	110	6	5.5	62	9	14.5	6	3	50.0	50.0	50
V. pilosa	78	72	7	9.7	25	10	40.0	4	3	25.0	75.0	25
V. khandalensis	85	78	6	7.7	32	6	18.8	3	1	66.7	33.3	66.7

The highest crossed-seed germination (79.2%) was recorded in *V. radiata* var. *sublobata* crosses, while the lowest germination (12.5%) was observed in *V. bournia* crosses. A total of 170 plants survived in *V. radiata* var. *sublobata*, whereas only three plants survived in *V. khandalensis*. At harvest, 152 plants were productive in *V. radiata* var. *sublobata*, while only one productive plant was recorded in *V. khandalensis*.

The crossability percentage ranged from 5.5 to 39.1. A crossability of 39.1% was observed in *V. radiata* with wild *V. radiata* var. *sublobata* species and 5.5% was observed in *V. capensis*. Hybrid breakdown of 89.4% was observed in the cross *V. radiata* × *V. radiata* var. *sublobata*, while the lowest, 33.3%, occurred in the cross *V. radiata* × *V. khandalensis*. Hybrid sterility was 66.7% in *V. radiata* × *V. khandalensis* and 10.6% in *V. radiata* × *V. radiata* var. *sublobata*. The percentage of lethality among crosses ranged from 10.6 in *Vigna radiata* var. *sublobata* to 66.7% in *Vigna radiata* × *V. khandalensis*.

Table 3a and 3b represents the morphological characters of the parents of *Vigna* species, and Table 4a and 4b represents the morphological characters of F_1 hybrids of *Vigna* species with *Vigna radiata*. Plates 1 to 4 illustrate these observations.

6.2.1.3 Discussion

Vigna species crosses studies have been carried out by many workers in the early 1960s, especially mungbean crossability with their wild progenitors (Dana 1964; De and Krishnan 1966; Chowdhary and Chowdhari 1974; Biswas and Dana 1975; Karmarkar and Dana 1987). A maximum hybrid sterility of 66.7% was recorded in *V. radiata* × *V. khandalensis* crosses and a minimum of 10.6% in *V. radiata* × *V. radiata* var. *sublobata*. A maximum hybrid breakdown of 89.4% was observed in *Vigna radiata* x *V. radiata* var *sublobata*) and a minimum of 33.3% in *V. radiata* × *V. khanalensis*. The lethality percentage of wide crosses ranged from 10.6% to 66.7%, with the maximum observed in *V. radiata* × *V. khandalensis* and the minimum of 10.6% recorded in *V. radiata* var. *sublobata*. Similar results were reported in wild *Vigna* accessions by Al-Yasiri and Coryne (1966), Chen et al. (1989), Adinarayanamurthy et al. (1993), Ganeshram (1993), and Renganayaki (1985). Stebbins (1958) attributed hybrid weakness, reduced viability, lethality, and sterility as natural mechanisms for maintaining the integrity of related species. Among the different *Vigna* species, *V. radiata* var. *sublobata* is amenable for interspecific crosses development and also closely related to *V. radiata*. Similarly, interspecific crosses between mungbean and its wild relatives was successfully done by many workers such as Renganayaki (1985), Pandae et al. (1990), Siriwardhane et al. (1991), Ganeshram (1993), Singh et al. (1996), Subramanian and Muthiah (2000), Uma maheshwari (2002), Pandiyan et al. (2010a, b), Bhanu et al. (2017), Lekhi et al. (2018), Pandiyan et al. (2006), Pandiyan et al. (2007), (2010a, b), and (2012a). The interspecific crosses of mungbean × ricebean, blackgram × adzuki bean, mungbean × *Vigna radiata* var. *sublobata*, and ricebean × adzuki bean were successfully performed by Bhanu et al. (2017). Bhanu et al. (2017) developed fertile hybrids in greengram variety TM96–2 crossed with ricebean genotype RB1. They also found that the interspecific crosses showed less germination (Lekhi et al. 2018). F1s

Table 3a Morphological characters of parents of *Vigna* species

S. No.	Particulars	Mungbean VRM (GG) 1	*Vigna sublobata*	*Vigna mungo* var. *silvestris*	Ricebean (TNAU Red) *Vigna umbellata*	*Vigna hainiana*	*Vigna trilobata*	*Vigna acconitifolia*	*Vigna settulosa*	*Vigna stipulacea*
1	Plant growth habitat	Tendril absence	Tendrils presence	Partial tendrils	Tendrils presence	Tendril absence	Tendrils	Tendrils	Tendrils presence	No Tendrils
2	Stem color	Green	Green	Purple red	Brown	Green	Green pink	Green	Green	Green
3	Leaves shape	Ovate	Ovate	Ovate lanceolate	lanceolate	Ovate	Trifoliolate	Ovate	Lanceolate Ovate	Ovate
4	Leaf color	Dark green	Green	Green, pale green	Pale green	Green	Green	Green	Pale green	Green
5	Leaf pubescence	Dense	Hairy	Hairy, glabrous	Glabrous	Glabrous	Slightly pubescence	Slightly pubescence	Slightly pubescence	Slightly pubescence
6	Flower color	Pale yellow	Yellowish green	Yellow	Bright yellow	Yellowish green	Yellow	Yellow	Pale yellow	Pale Yellow
7	Peduncle color	Green	Green	Reddish purple	Brown	Green	Green	Green	Brown	Green
8	Number of flowers per cluster	10 flowers	10–15 flowers	10–15	10–15 flowers One to three	8–10	10–15	10	10–15 flowers One to three	5–8
9	No. of peduncle per leaf axil	One	One	One	One to three	One	one	One	One Three	one
10	Pod pubescence	Dense	Glabrous	Slightly hairy	Glabrous	Slightly hairy	Slightly pubescence	Glabrous	Slightly pubescence	Slightly pubescence
11	Pod color (immature)	Green	Black	Brownish black	Black	Black	Black	Black brown and straw	Black	Black
12	Pod shape	Straight	Curved	Straight	Curved	Straight	Straight	Straight	Straight	Straight

13	No of seeds per pod	10	6	5	6	8	10	6	10	7
14	100 seed weight	3.38 g	4.98 g	2.8–3.5	4.98 g	3.2	1.0–2.0	2.5	2.5	1.8
15	Seed coat colure	Green	Brown	Brown black	Brown	Brown	Black	Brown, black	Green	Green
16	Seed coat texture	Smooth	Shiny	Rough	Shiny	Rough	Shiny	Shiny	Shiny	Shiny
17	Seed shape	Oval	Reniform	oval	Reniform	Oval	Oval	Oval	Reniform	Oval

Table 3b Morphological characters of parents of *Vigna* species

S. No.	Particulars	*Vigna angularis*	*Vigna vexillata*	*Vigna glabrascense*	*Vigna marina*	*Vigna bournia*	*Vigna capensis*	*Vigna pilosa*	*Vigna khandalensis*
1	Plant growth habitat	Tendrils	Tendril presence	Tendrils presence	No tendrils	Tendril presence	Tendrils present	Tendrils	No tendrils
2	Stem color	Green	Green pink	Green	Purple red	Green	brown	Green	Pale green
3	Leaves shape	Ovate	Lanceolate Ovate	Ovate	Ovate	Ovate	Ovate	Lanceolate	Lanceolate Ovate
4	Leaf color	Green	Green	Green	Green, pale green	Green	Dark green	Pale green	Pale green
5	Leaf pubescence	Dense pubescence	Dense	Hairy	glabrous	Highly pubescence	Moderate	Slightly pubescence	Moderate
6	Flower color	Yellow	Purple yellow	Yellowish green	Yellow Purple	Yellow	Bright yellow	Purple	Bright yellow
7	Peduncle color	Green	Green	Green	Green	Green	Brown	Pale green	Green
8	Number of flowers per cluster	10	5–8 flowers	10–15 flowers	5–8	8–10	3–5 flowers	5	3–5 flowers
9	No. of peduncle per leaf axil	One	One	One	One	One	one	One	one
10	Pod pubescence	Pubescence	Dense	Glabrous	Glabrous	Dense hairy	Moderate	Velvet pubescence	Moderate
11	Pod color (immature)	Black brown and straw	Straw	Black	Brownish black	Black	Straw brown	Black	Straw brown
12	Pod shape	Straight	Straight	Straight	Straight	Straight	Curved	Straight	Straight
13	No of seeds per pod	6	12	5	5	10	7	5	5
14	100 seed weight	2.5	4.5	3.5	2.8–3.5	2.5	2.80 g	3.2	2.5 g

15	Seed coat colure	Brown, black	Green, Yellow	Brown	Brown black	Brownish black	Mixture of brown, green and white dots	Green	Green
16	Seed coat texture	Shiny	Smooth	Shiny	Rough	Rough	Shiny	Shiny	Shiny
17	Seed shape	Kidney	Oval	Reniform	oval		Oval	Reniform	Oval

Table 4a Morphological characters of F1 of *Vigna* species with *Vigna radiata*

S. No.	Particulars	Mungbean VRM (GG) 1	*Vigna radiata* × *Vigna. radiata* var. *sublobata*	*Vigna radiata* × *V. mungo* var. *silvestris*	*Vigna radiata* × *Vigna hainiana*	*Vigna radiata* × *Vigna umbellata*	*Vigna radiata* × *Vigna trilobata*	*Vigna radiata* × *Vigna acconitifolia*	*Vigna radiata* × *Vigna settulosa*	*Vigna radiata* × *Vigna stipulacea*
1	Plant growth habitat	Tendril absence	Tendrils presence	Tendrils presence	Tendril absence	Tendrils presence	Tendrils	Tendrils	Tendrils presence	No tendrils
2	Stem color	Green	Green	Greenish purple	Green	Green	Green pink	Green	Green	Green
3	Leaves shape	Ovate	Ovate	Ovate	Ovate	Lanceolate	Ovate	Ovate	Lanceolate Ovate	Ovate
4	Leaf color	Dark green	Dark green	Greenish purple	Green	Pale green	Green	Green	Green	Green
5	Leaf pubescence	Dense	Dense hairy	Dense	Glabrous	Glabrous	Dense pubescence	Slightly pubescence	Slightly pubescence	Slightly pubescence
6	Flower color	Pale yellow	Pale yellowish green	Pale yellow	Pale yellowish	Bright yellow	Yellow	Yellow	Greenish yellow	Yellow
7	Peduncle color	Green	Green	Greenish purple	Green	Green	Green	Green	Green	Green
8	Number of flowers per cluster	10 flowers	10–20 flowers	10–15	8	10–20 flowers One to three	5–8	8	10–15 flowers	5
9	No. of peduncle per leaf axil	One	One	One	One	One - three	One	One	One	One
10	Pod pubescence	Dense	Dense	Dense hairy	Slightly hairy	Glabrous	Slightly pubescence	Slightly Glabrous	Slightly pubescence	Slightly pubescence
11	Pod color (immature)	Green	Green	Brownish black	Black	Yellow	Black	Black brown and straw	Black	Greenish Yellow

12	Pod shape	Straight	Straight	Straight	Straight	Curved	Straight	Straight	Straight	Straight
13	No of seeds per pod	10	8	8	6	8	8	6	8	5
14	100 seed weight	3.38 g	2.9	2.5	2.8	2.5	1.0–2.0	2.5	2.6	1.5
15	Seed coat color	Green	Brown/ black	Brown black	Brown	Green	Green	Green	Green	Green
16	Seed coat texture	Smooth	Smooth	Rough	Rough	Shiny	Smooth	Smooth	Shiny	Shiny
17	Seed shape	Oval	oval	oval	Round	Reniform	Round	Round	Reniform	Oval

Table 4b Morphological characters of F1 of *Vigna* species with *Vigna radiata*

S. No.	Particulars	*Vigna radiata* × *V. angularis*	*Vigna radiata* × *Vigna Vexillata*	*Vigna radiata* × *Vigna glabrascense*	*Vigna radiata* × *Vigna marina*	*Vigna radiata* × *Vigna bournia*	*Vigna radiata* × *Vigna capensis*	*Vigna radiata* × *Vigna pilosa*	*Vigna radiata* × *Vigna khandalensis*
1	Plant growth habitat	Tendrils	Tendril presence	Tendrils presence	No tendrils	Tendril presence	Tendrils present	Tendrils	No Tendrils
2	Stem color	Green	Green pink	Green	Green	Green	Brown	Green	Green
3	Leaves shape	Ovate	Lanceolate Ovate	Ovate	Ovate	Ovate	Ovate Lanceolate	Ovate	Ovate
4	Leaf color	Green	Pale Green	Green	Green	Green	Green	Green	Green
5	Leaf pubescence	Slight pubescence	Dense	Glabrous	glabrous	Highly pubescence	Moderate	Slightly pubescence	Moderate
6	Flower color	Yellow	Yellow	Yellow	Yellow	Yellow	Yellow	Yellow	Greenish yellow
7	Peduncle color	Green	Green	Green	Green	Green	Green	Green	Green
8	Number of flowers per cluster	5	10 flowers	10–15 flowers	5–8	5–8	5–8 flowers	5	5 flowers
9	No. of peduncle per leaf axil	One	One	One	One	One	one	One	one
10	Pod pubescence	Slightly Pubescence	Dense	Slightly Glabrous	Slightly Glabrous	Dense hairy	Moderate	Slightly pubescence	Moderate
11	Pod color (immature)	Black	Black	Black	black	Black	Black	Black	Greenish yellow
12	Pod shape	Straight	Straight	Straight	Straight	Straight	Straight	Straight	Straight

13	No. of seeds per pod	5	5	6	5	8	10	5	5
14	100 seed weight	2.0	2.6	2.8	2.8	2.5	2.5	2.2	2.5 g
15	Seed coat colure	Green	Green,	Green	Black	Black	Black brown	Green	Green
16	Seed coat texture	Shiny	Smooth	Shiny	Rough	Rough	Shiny	Shiny	Shiny
17	Seed shape	Oval	Oval	Reniform	oval	oval	Oval	Reniform	Oval

reaction showed the transfer of mungbean yellow mosaic virus gene from *Vigna mungo* to *Vigna radiata* (Chitra et al. 2018). Pre-fertilization barriers occurred when mungbean was pollinated with pollen from Vigna stipulacea (used as the male parent), whereas fertile hybrids were recovered when mungbean was used as the female parent Singh et al. (2020) and Prithviraj and Niranjana Murthy (2018); mungbean and ricebean by Sharma et al. (2022); mungbean and adzuki bean by Ahn and Hartmann (1978a); and *Vigna* species crossability by Audilakshmi and Chandel (1990), Chen et al. (1983), and Dana (1966). Interspecific hybridization has been explored among various *Vigna* species. Gosal and Bajaj (1983) reported that hybridization between *Vigna mungo* and *Vigna radiata* (Gupta et al. 2002). Hybrid between blackgram x greengram derivatives and adzuki bean Pal et al. (2005). Crosses between *Vigna raiata x Vigna mungo* were reported by Subramanian and Muthiah (2001), and additional interspecific cross combinations among *Vigna* species were documented by (Sudha et al. 2013).

6.2.1.4 Conclusion

Vigna species play an important role as protein-rich pulse crops. They have high economic value due to their nutritional quality. Interspecific hybridization provides a way to transfer useful genes from wild species to cultivated types. Among the *Vigna* species, *V. radiata* shows wide compatibility and can cross with many species, though the level of crossability varies among cross combinations. Even F1s showed hybrid inviability, hybrid sterility, and hybrid lethality. The broad genetic base also increased through interpecific hybridization.

References

Adinarayanamurty VV, Rao MVB, Satyanarayana A, Subramanyam D (1993) The crossability of *V. mungo* and *V. radiata* with *V. trilobata*. Int J Trop Agric 11:209–213

Ahn CS, Hartmann RW (1978a) Inter-specific hybridization between Mungbean (*V. radiata* (L.) Wilczek) and Adzukibean (*V. angularis* (wild) ohwi and ohashi). J Am Soc Hortic Sci 103:3–6. https://doi.org/10.21273/JASHS.103.4.435

Ahn CS, Hartmann RW (1978b) Interspecific hybridization among four species of the genus *Vigna* Savi. In: Report of first international Mungbean symposium. Univ. Philippines, Los Banios, pp 240–246

Ahuja MR, Singh BV (1977) Induced genetic variability in Mungbean through inter-specific hybridization. Indian J Genet Plant Breed 37:133–136

Al-Yasiri SA, Coryne DP (1966) Interspecific hybridization in the genus *Phaseolus*. Crop Sci 6:59–60. https://doi.org/10.2135/cropsci1966.0011183X000600010018x

Asthana S, Singh A, Singh NP, Ashthana AN (1994) Biotechnological strategies to combat biotic stresses in pulse crops. In: Proc. Int. Symp. Pulses Res., New Delhi, p 269

Audilakshmi B, Chandel KPS (1990) Interspecific hybridization and crossability studies among cultivate and closely related wild *Vigna* species in Indian Gene Centre. Indian J Plant Genet Resour 3:46–60

Baudoin JP, Marechal R (1988) Taxonomy and evolution of the genus *Vigna*: Mungbean. In: Proceedings of second international symposium. AVRDC, Bangkok, Shanhua, Taiwan

Baudoin JP, Marechal R (1989) Taxonomy and evolution of the genus *Vigna*. Mungbean, 1

Bhanu AK, Kumar P, Singh MN, Srivastava K, Hemantaranjan A (2017) Assessment of genetic purity of inter-specific F_1 hybrids involving *Vigna radiata* and *Vigna umbellata*. J Exp Biol Agric Sci 5(5):637–643

Biswas MR, Dana S (1975) Blackgram x Rice bean cross. Cytologia 40:787–795. https://doi.org/10.1508/cytologia.40.787

Boling M, Sander DA, Matlock RS (1961) Mungbean hybridization technique. Agron J 53:54–55

Chandel KP, Laster RN (1991) Origin and evolution of Asiatic *Vigna* species. In: Golden jubilee celebration symposium on grain legumes. IARI, New Delhi, pp 25–45

Chandel KPS, Lester RN, Starling RJ (1984) The wild ancestors of urd and Mungbeans [*Vigna mungo* (L.) Hepper and *V. radiata* (L.) Wilczek]. Bot J Linn Soc 89:85–96. https://doi.org/10.1111/j.1095-8339.1984.tb01002.x

Chen NC, Baker RL, Honma S (1983) Interspecific crossability among four species of *Vigna* food legumes. Euphytica 32:925–937. https://doi.org/10.1007/BF00042175

Chen HK, Mok MC, Shanmugasundaram S, Mok DW (1989) Interspecific hybridization between *Vigna radiata* (L.) Wilczek and *V. glabrescens*. Theor Appl Genet 78(5):641–647. https://doi.org/10.1007/BF00262558

Chitra S, Kalaimagal T, Thirumurugan T (2018) Development of MYMV resistant prebreeding lines in mungbean (*Vigna radiata*) through Interspecific hybridization. Int J Curr Microbiol App Sci 7(01):2674–2677

Chowdhury RK, Chowdhury JB (1974) Induced amphidiploidy in Phaseolus aureus Roxb.x *P. mungo*L. hybrids. Crop Improv 1:46–52

Dana S (1964) Interspecific cross between tetraploid *Phaseolus* species and *P.ricciardianus* Ten. Nucleus 7:1–10

Dana S (1966) Interspecific hybrid between *Phaseolusmungo* L. × *Phaseolustrilobus* Ait. J Cytol Cytogenet 1:61–66

Dana S, Karmakar PG (1990) Species relation in *Vigna* subgenus *Ceratotropis* and its implications in breeding. Plant Breed Rev 8:19–42. https://doi.org/10.1002/9781118061053

De DN, Krishnan R (1966) Cytological studies of the hybrid *Phaseolus aureus* and *P. mungo*. Genetica 37:588–600

Delgado-Salinas A, Lavin M, Snak C, Lewis GP (2022) Systematics of *Vigna* subgenus Lasiospron (Leguminosae:Papilionoideae:Phaseolinae). Syst Bot 47(1):97–124. https://doi.org/10.1600/036364422X16442668423428

Ganeshram S (1993) Evaluation of some genotypes interspecific hybrids and derivatives of greengram (*V. radiata* (L.) Wilczek x Black gram (*Vigna mungo* (L.) Hepper) crosses. M.Sc. (Ag.) Thesis, Tamil Nadu Agricultural University, Coimbatore

Gopinathan MC, Babu CR (1986) Aunique growth pattern associated with cleistogamyina tropical legume, *Vigna minima* Roxb. Ohwi and Ohashi Leguminosae. Bot J Linn Soc 92:263–268. https://doi.org/10.1111/j.1095-8339.1986.tb01431.x

Gosal SS, Bajaj YPS (1983) Interspecific hybridization between *Vigna mungo* and *Vigna radiata* through embryo culture. Euphytica 32:129–137. https://doi.org/10.1007/BF00036873

Gupta VP, Plaha P, Rathore PK (2002) Partiallyfertile interspecific hybridsbetweena black gram × green gram derivative and adzuki bean. Plant Breed 121:182–183. https://doi.org/10.1046/j.1439-0523.2002.00675.x

Harlan JR, de Wet JMJ (1971) Towards a rational classification of cultivated plants. Taxon 20:509–517. https://doi.org/10.2307/1218252

Kaga A, Isemura T, Tomooka N, Vaughan DA (2008) The genetics of domestication of the azuki bean. Geneticsm 178:1013–1036. https://doi.org/10.1534/genetics.107.078451

Karmarkar PG, Dana S (1987) Cytogenetic identification of a *Vigna sublobata* collection. Nucleus 30:47–50

Kumar S, Gupta S, Chandra S, Singh BB (2004) How wide is the genetic base of pulse crops? In: Ali M, Singh BB, Kumar S, Dhar V (eds) Pulses in new perspective. Indian Society of Pulses Research and Development, Kanpur, pp 211–221

Lawn RJ, Cottell A (1988) Wild Mungbean and its relative in Australia. Biologist 35:267–273

Lekhi P, Gill RK, Kaur S, Bains TS (2018) Generation of interspecific hybrids for introgression of Mungbean yellow mosaic virus resistance in [*Vigna radiata* (L.)Wilczek]. Legume Res 41(4):526–531. https://doi.org/10.18805/LR-3808

Marechal R, Mascherpa JM, Stainer F (1978) Etude taxonomique d'un groupe complexed' speces des genres *Phaseolus*et *Vigna* (Papilionaceae) sur la base de donnees morphologiques et polliniques, traitees par l'analyse informatique. Boissiera 28:177–178

Pal SS, Sandhu JS, Inderjit Singh IS (2005) Exploitation of genetic variability in interspecific cross between *Vigna mungo* × *V. umbellata*. Indian J Pulses Res 18:9–11

Pandae K, Raghavanshi SS, Prakesh P (1990) Induced high yielding amphiploid of *Vigna radiata* x *Vigna mungo*. Cytologia 55:249–253. https://doi.org/10.1508/cytologia.55.249

Pandiyan M, Subbalakshmi B, Jebaraj S (2006) Genetic variability in greengram (*Vigna radiata* (L) wilczek). Int J Plant Sci 1(l):72–75

Pandiyan M, Subbalakshmi B, Alice D, Marimuthu R (2007) Screening of Mungbean [*Vigna radiata* (L.) Wilczek] germplasm for Mungbean yellow mosaic virus. Plant Arch 7(1):375–376

Pandiyan M, Senthil N, Ramamoorthi N, Muthiah AR, Tomooka N, Duncan V, Jayaraj T (2010a) Interspecific hybridization of *Vigna radiata* × 13 wild *Vigna* species for developing MYMV donar. Electron J Plant Breed 1:600–610

Pandiyan M, Senthil N, Sivakumar P, Muthiah AR, Ramamoorthi N (2010b) Genetic diversity analysis among green gram genotypes using RAPD markers. Electron J Plant Breed 1(4):466–473

Pandiyan M, Senthil N, Suresh R, Chakravarthy N, Packiraj D, Jagadeesh S (2012a) Interspecific hybridization of *Vigna radiata* × *V. trilobata*. Euphytica 35:1017–1022

Pandiyan M, Senthil N, Packiaraj D, Gupta S, Nadarajan N, Thanga PR, Suresh R, Jagadeesh S (2012b) Characterisation and evaluation of 646 greengram (*Vigna radiata*) genotypes for constituting core collection. Wudpecker J Agric Res 1(8):294–301

Pratap A, Kumar J (2011) History origin and evolution. In: Pratap A, Kumar J (eds) Biology and breeding of food legumes. CABI, Oxfordshire

Prithviraj SK, Niranjana Murthy NM (2018) Interspecific hybridization between Mungbean (*Vigna radiata*) and rice bean (*Vigna umbellata*) for crossability and related parameters. Mysore J Agric Sci 52(2):316–322

Rawal KM (1975) Natural hybridization among wild, weedy and cultivated *Vigna unguiculata* L. Walp. Euphytica 24:699–707. https://doi.org/10.1007/BF00132908

Renganayaki K (1985) Studies on genetic differentiation between three species of Vigna Savi. M.Sc. (Ag.) Thesis, Tamil Nadu Agricultural University, Coimbatore

Sharma S, Mittal RK, Sood VK, Chaudhary HK (2022) Development of interspecific hybrids between urdbean & Mungbean. Environ Conserv J 23(1&2):315–321. https://doi.org/10.36953/ECJ.021892-2171

Singh HB, Joshi BS, Chandel KP, Pant KC, Saxena RK (1974) Genetic diversity in some Asiatic *Phaseolus* species and its conservation. Indian J Genet Plant Breed 34:52–57

Singh MN, Singh RM, Singh UP (1996) Studies on hybrids and transgressive segregates in wide crosses of Mungbean and urdbean. Indian J Genet Plant Breed 59:109–113

Singh SK, Singh MN, Choudhary VK, Singh MK, Tigga A (2020) Studies on the interspecific hybrids between Mungbean and urdbean. Int J Curr Microbiol App Sci 9(5):364–369

Siriwardhane D, Egawa Y, Tomooka N (1991) Cross-compatibility of cultivated adzuki bean (*Vigna angularis*) and rice bean (*V. umbellata*) with their wild relatives. Plant Breed 107:320–325. https://doi.org/10.1111/j.1439-0523.1991.tb00555.x

Smartt J (1985) Evolution of grain legumes. III. Pulses in the genus *Vigna*. Exp Agric 21:87–100. https://doi.org/10.1017/S0014479700012370

Somta P, Ammaranan C, Ooi PAC, Srinivas P (2007) Inheritance of seed resistance to bruchids in cultivated Mungbean (*Vigna radiata*, L. Wilczek). Euphytica 155:47–55. https://doi.org/10.1007/s10681-006-9299-9

Stebbins GL (1958) The in viability, weakness and sterility of interspecific hybrids. Adv Genet 9:147–215. https://doi.org/10.1016/S0065-2660(08)60162-5

Subramanian A, Muthiah AR (2000) Interspecific hybridization between *V. radiata* (L.) Wilczek and blackgram *V. mungo* (L.) Hepper. Legum Res 24(3):154–158

Subramanian A, Muthiah AR (2001) Interspecific hybridization between *Vigna radiata* (L.) Wilczek and *V. mungo* (L.) Hepper. Legume Res 24:154–158

Sudha M, Anusuya P, Nawkar GM, Karthikeyan A, Nagarajana P, Raveendrana M, Senthil N, Pandiyan M, Angappan K, Balasubramanian P (2013) Molecular studies on Mungbean (*Vigna radiata* (L.) Wilczek) and ricebean (*Vigna umbellata* (Thunb.)) interspecific hybridisation for Mungbean yellow mosaic virus resistance and development of species-specific SCAR marker for ricebean. Arch Phytopathol Plant Prot 46:503–517. https://doi.org/10.1080/03235408.2012.745055

Uma Maheswari D (2002) Wide hybridization in the genus *Vigna*. M.Sc. (Ag.) Thesis, TNAU, Coimbatore

Vaughan DA, Tomooka N, Kaga A (2005) Azuki bean. In: Singh RJ, Jauhar PP (eds) Genetic resources, chromosome engineering and crop improvement. Grain legumes, vol I. CRC Press, Boca Raton, FL, pp 341–353

Vavilov NI (1926) Studies on the Origin of Cultivated Plants. Institut de Botanique Appliquée et d'Amélioration des Plantes

Verma RP, Singh DP (1986) Problems and prospects of interspecific hybridization involving greengram and blackgram. Indian J Genet 37:130–132

Yamaguchi H (1992) Wild and weed azuki beans in Japan. Econ Bot 46(4):384–394. https://doi.org/10.1007/BF02866509

Zukovskij PM (1962) Cultivated plants and their wild relatives. Commonwealth Agriculture Bureau

Molecular Breeding Approaches for Enhancing Pest Resistance in Mungbean (*Vigna radiata* L. Wilczek)

M. Ravi, A. Sowmiya, G. Anand, V. Radhakrishnan,
C. Pradipa, and P. Priyadharshini

Abstract

Mungbean (*Vigna radiata* L. Wilczek), a nutrient-rich legume, is a critical crop in Asia, Africa, and Latin America. Its production is frequently threatened by pests such as bruchids, whiteflies, aphids, and pod borers, resulting in significant yield losses. Conventional breeding for pest resistance has been slow due to the complex genetics of resistance traits, limited genetic diversity, and prolonged breeding cycles. Molecular breeding, leveraging advances in genomics, marker-assisted selection (MAS), genetic engineering, and genome editing, has revolutionized pest resistance breeding in mungbean. This chapter discusses the molecular mechanisms of pest resistance, highlights of advancements in genomics-based breeding strategies, and outlines of future directions to address challenges in pest-resistant mungbean varietal development.

Keywords

Breeding · Pest · Mungbean · Resistance · Genome · Legume

M. Ravi · G. Anand
ICAR-KrishiVigyan Kendra, Sandhiyur, Salem, India

A. Sowmiya
Senior Research Fellow, ICAR-KrishiVigyan Kendra, Sandhiyur, Salem, India

V. Radhakrishnan (✉)
AnbilDharmalingam Agricultural College and Research Institute, NavalurKuttapattu, Tiruchirappalli, India

C. Pradipa
Agro Climatic Research Centre, Tamil Nadu Agricultural University, Coimbatore, India

P. Priyadharshini
Forest College and Research Institute, Mettupalayam, India
e-mail: priyadharshini.p@tnau.ac.in

S. Paramasivam et al. (eds.), *Mungbean*,
https://doi.org/10.1007/978-981-95-7637-1_4

1 Introduction

Mungbean (*Vigna radiata* L. Wilczek) is a short-duration legume crop widely cultivated in Asia, Africa, and Latin America, contributing significantly to food and nutritional security. Known as the "poor man's meat," it is a rich source of protein, vitamins, minerals (particularly iron and zinc), and dietary fiber, making it essential for combating malnutrition, particularly in developing countries (Nair et al. 2019). Mungbean is also highly valued for its agronomic benefits, including its ability to fix atmospheric nitrogen in symbiosis with *Rhizobium* bacteria, thus improving soil fertility and reducing the dependency on synthetic nitrogen fertilizers (Praharaj and Maitra 2020; Kebede 2021).

Despite its significance, mungbean production faces considerable challenges, primarily due to biotic and abiotic stresses. Among these, biotic stresses caused by insect pests are a major constraint. Globally, pests such as bruchids (*Callosobruchus* spp.), whitefly (*Bemis iatabaci* Gennadius), aphids (*Aphis craccivora* Koch.), and pod borers (*Maruca vitrata* Fabricius) cause yield losses of 20–50% annually (Lal 1985). These losses not only impact the livelihood of farmers but also hinder efforts to achieve sustainable production to meet growing global demand.

Traditional breeding efforts have achieved some success in developing pest-resistant mungbean varieties; however, the complex genetic basis of resistance traits, coupled with environmental variability, poses significant challenges. Conventional approaches often involve long breeding cycles and are constrained by the lack of precise phenotyping tools. In this context, molecular breeding has emerged as a transformative approach, offering precise, efficient, and reliable tools to enhance pest resistance in mungbean.

2 Challenges in Pest Management in Mungbean

Mungbean faces several challenges in pest management, which complicate efforts to develop effective resistance strategies. First, pest diversity is a major issue, as mungbean is susceptible to a wide range of pests throughout its life cycle, from seedling to storage. Each pest has distinct biology and interactions with the host plant, requiring tailored management approaches (Sehgal and Ujagir 1988). Additionally, environmental factors play a significant role in pest dynamics, with varying climatic conditions affecting pest populations and making resistance breeding both complex and site-specific (Dogimont et al. 2010). The rapid adaptation of pests further complicates management, as pests evolve quickly and may overcome the resistance mechanisms introduced through traditional breeding methods (Varshney et al. 2018). Lastly, limited resources in terms of genomic data and identified resistance genes in mungbean hinder the progress of breeding pest-resistant varieties, especially when compared to major cereal crops with well-established genetic resources (Guo 2024). These challenges highlight the need for more advanced breeding techniques and sustainable pest management solutions to ensure the continued productivity and sustainability of mungbean cultivation.

3 Limitations of Conventional Breeding

Conventional breeding has been the primary method for developing pest-resistant mungbean varieties. However, the progress has been limited due to several factors:

1. *Complexity of resistance traits*: Resistance to many pests involves quantitative traits controlled by multiple genes, making it difficult to select and breed effectively (Bohra et al. 2014).
2. *Pest Adaptation*: The rapid evolution of pests often overcomes resistance mechanisms introduced through traditional breeding (Varshney et al. 2018).
3. *Limited Genetic Diversity*: Narrow genetic diversity in mungbean germplasm restricts the availability of resistance sources for breeding programs (Dogimont et al. 2010).
4. *Long Breeding Cycles*: Traditional breeding methods are time-consuming and labor-intensive, delaying the development of resistant varieties (Rubiales et al. 2015).

4 Role of Molecular Breeding

Molecular breeding, an advanced approach integrating genomics, biotechnology, and breeding, offers precise tools to overcome these challenges. By utilizing molecular markers, high-throughput sequencing technologies, and genome editing, molecular breeding accelerates the development of pest-resistant mungbean varieties (Varshney et al. 2018). The integration of molecular breeding approaches enables:

1. *Precise Trait Mapping*: Identification and manipulation of genes or quantitative trait loci (QTL) associated with pest resistance.
2. *Marker-Assisted Selection (MAS)*: Efficient selection of resistant lines, reducing dependency on phenotypic screening.
3. *Genome Editing*: Precise modification of susceptibility genes to enhance resistance without compromising other traits (Wu et al. 2022).
4. *Pyramiding Resistance Genes*: Combining multiple resistance genes into a single variety for durable resistance.

5 Importance of Molecular Breeding in Pest Resistance

Molecular breeding leverages genomic tools to identify and incorporate resistance genes into elite mungbean cultivars. Compared to conventional methods, molecular approaches offer several advantages:

- *Precision*: Genetic markers and genome editing techniques enable targeted introgression of resistance traits.

- *Efficiency*: Marker-assisted selection (MAS) and genomic selection (GS) reduce breeding cycles.
- *Durability*: Stacking multiple resistance genes ensures long-term effectiveness against pest adaptation.

6 Major Pests of Mungbean

Mungbean cultivation is significantly constrained by pest infestations, which result in yield losses during both the growing season and storage. Understanding the biology, behavior, and damage caused by these pests is essential for developing effective resistance breeding strategies.

6.1 Bruchids (*Callosobruchus* spp.)

Bruchids, commonly known as seed weevils, are the most destructive post-harvest pests of Mungbean. The most prevalent species include *Callosobruchus chinensis* L. and *Callosobruchus maculatus* Fab. These beetles are major storage pests that infest mungbean seeds during storage that significantly reduce seed viability and quality causing considerable economic losses.

6.1.1 Biology and Behavior

Female bruchids lay eggs on the surface of stored seeds. Upon hatching, the larvae penetrate the seeds, feeding on the internal tissues and reducing seed viability and weight. The larval feeding results in the production of hollowed-out seeds often filled with frass. Adults emerge from the seeds by creating circular exit holes.

6.1.2 Impact

Infestation reduces the nutritional and market value of mungbean seeds. In severe cases, the seeds are rendered unfit for consumption or planting. Infestation rates can escalate under warm and humid storage conditions, leading to significant losses.

6.1.3 Breeding Challenges

Bruchid resistance is often governed by polygenic traits, and screening for resistance in mungbean germplasm is resource-intensive. Molecular breeding tools, such as marker-assisted selection (MAS), have been instrumental in identifying resistance loci, such as those linked to seed coat hardness and biochemical defense mechanisms like protease inhibitors (Lin et al. 2023).

6.1.4 Breeding Target

Resistance has been linked to loci such as qBru1, which has been identified through quantitative trait loci (QTL) mapping. The integration of these loci into elite varieties using MAS has shown promising results (Wu et al. 2022).

6.2 Whitefly (*Bemisia Tabaci*)

Whiteflies are sap-sucking insects that feed on the undersides of mungbean leaves, causing direct and indirect damage. They are also the primary vectors of mungbean yellow mosaic virus (MYMV), the most devastating disease affecting mungbean cultivation.

6.2.1 Biology and Behavior

Adult whiteflies are small, winged insects that lay eggs on the underside of leaves. The nymphs and adults extract phloem sap, leading to physiological stress in plants. Infestation is often identified by yellowing and wilting of leaves, along with the presence of sooty mold due to honeydew excretion.

6.2.2 Impact of MYMV

MYMV infection can lead to stunted growth, chlorotic leaf symptoms, and severe yield losses, with reports of up to 100% crop failure under epidemic conditions. MYMV management through whitefly control is challenging due to the pest's high reproductive rate and resistance to insecticides (Lal 1985).

6.2.3 Breeding Challenges

Resistance to MYMV in mungbean involves complex gene interactions. Molecular approaches, such as QTL mapping and the identification of resistance genes (VrMYMIV-1 and VrMYMIV-2), are crucial for breeding efforts (Varshney et al. 2018).

6.2.4 Breeding Target

The resistance gene VrMYMIV-1, identified through QTL mapping, has been incorporated into MYMV-resistant varieties such as 'ML 267' and 'Meha' (Mathivathana et al. 2019).

6.3 Aphids (*Aphis craccivora)*

Aphids are another significant pest of mungbean, causing direct damage through sap feeding and indirect damage by transmitting viral diseases.

6.3.1 Biology and Behavior

Aphids are small, soft-bodied insects that reproduce rapidly, often forming large colonies on young shoots and leaves. They feed on phloem sap, which weakens plants and disrupts nutrient flow.

6.3.2 Impact

Infested plants exhibit leaf curling, chlorosis, and stunted growth. Aphids are also vectors for several viral diseases, compounding the damage. Their rapid population buildup under favorable conditions makes them particularly challenging to control.

6.3.3 Breeding Challenges

Resistance to aphids in mungbean is primarily associated with antixenosis (deterrence of aphid feeding and colonization) and antibiosis (negative effects on aphid reproduction). However, these traits are influenced by environmental factors, making breeding for resistance complex (Dogimont et al. 2010).

6.3.4 Breeding Target

Resistance mechanisms in mungbean, such as antibiosis and antixenosis, are being explored through transcriptome analysis to identify key resistance genes for breeding programs (Sodedji et al. 2020).

6.4 Pod Borers (*Maruca Vitrata*)

Pod borers are among the most destructive pests during the reproductive stage of mungbean, directly impacting pod and seed quality. Pod borers damage developing pods, leading to significant yield losses.

6.4.1 Biology and Behavior

Maruca vitrata larvae bore into flower buds, flowers, and developing pods, feeding on internal tissues. The pest has a broad host range, allowing it to sustain populations across cropping seasons. The adult moths are migratory, which contributes to widespread infestations.

6.4.2 Impact

Infestation results in flower and pod abortion, malformed pods, and reduced seed quality. Yield losses due to pod borer damage can range from 30% to 50% under severe infestations (Khan et al. 2022).

6.4.3 Breeding Challenges

Resistance to pod borers in mungbean is governed by traits such as pod wall thickness and the production of secondary metabolites. These traits are often quantitative, requiring advanced molecular tools like genomic selection and genome editing for effective incorporation into breeding programs.

6.4.4 Breeding Target

While MAS for pod borer resistance in mungbean is in its early stages, candidate resistance markers identified in related legumes are being tested for transfer into mungbean (Cheema et al. 2017).

7 Integrating Pest Biology and Resistance Breeding

A thorough understanding of the biology and behavior of major pests is essential for identifying resistance mechanisms and breeding strategies. Molecular tools, such as RNA sequencing and transcriptomic analysis, can uncover pest-responsive genes and pathways, aiding in the development of durable resistance. Future breeding programs must focus on combining pest resistance with other desirable traits, such as yield and stress tolerance, to ensure sustainable mungbean production.

8 Molecular Mechanisms of Pest Resistance

Resistance in mungbean against pests is mediated by several mechanisms at physiological, biochemical, and molecular levels, broadly categorized as antibiosis, antixenosis, and tolerance. These mechanisms are controlled by complex genetic networks and signaling pathways, making them crucial targets for molecular breeding strategies.

8.1 Antibiosis

Antibiosis involves the production of plant traits or compounds that negatively affect pest survival, growth, and reproduction. This defense mechanism is particularly effective against bruchids and pod borers.

8.1.1 Secondary Metabolites

Secondary metabolites play a vital role in mungbean resistance:

1. *Phenolic Compounds*: These interfere with pest digestion by inhibiting enzymes, thereby reducing bruchid larval growth (Kpoviessi et al. 2021).
2. *Protease Inhibitors*: Protease inhibitors disrupt the protein digestion process in pests, leading to reduced larval development.
3. *Flavonoids and Isoflavonoids*: These function as both antifeedants and reproductive disruptors, significantly impairing pest population dynamics (Sodedji et al. 2020).

8.1.2 Defensive Proteins

Mungbean synthesizes specific proteins in response to pest attack:

1. *Chitinases*: These degrade the chitin in the exoskeleton of insects, making them highly effective against bruchids and aphids (Rodríguez-Sifuentes et al. 2020).
2. *α-Amylase Inhibitors*: These proteins inhibit carbohydrate digestion, starving the pests (Sharma et al. 2008).

3. *Molecular pathways*: Jasmonic acid (JA) and salicylic acid (SA) signaling pathways regulate the biosynthesis of defensive metabolites and proteins (Li et al. 2019). JA is particularly significant in herbivory-induced defenses.

8.2 Antixenosis

Antixenosis involves plant traits that deter pests from feeding, oviposition, or colonization. These traits make the plant unattractive or unsuitable for pests.

8.2.1 Physical Traits

Structural barriers play an essential role in deterring pests:

1. *Trichomes*: Trichome density on mungbean leaves and pods acts as a physical barrier, discouraging feeding and egg-laying (Soundararajan et al. 2013). Glandular trichomes release toxic compounds that further deter pests.
2. *Pod Hardness*: A thick and tough pod wall provides resistance against bruchids by preventing larval penetration (War et al. 2018).

8.2.2 Chemical Traits

1. *Volatile Organic Compounds (VOCs)*: VOCs act as deterrents or attract natural enemies of pests, aiding in biocontrol (War et al. 2020).
2. *Saponins and alkaloids*: These compounds reduce pest feeding and oviposition, significantly lowering pest pressure (Divekar et al. 2023).

8.2.3 Molecular Pathways

Ethylene signaling, in conjunction with JA, regulates the synthesis of VOCs and cuticle-related compounds, enhancing the plant's defense mechanisms (Pratap et al. 2021).

8.3 Tolerance

Tolerance enables mungbean plants to withstand pest damage without significant yield loss. This resistance mechanism relies on the plant's capacity to repair and sustain growth under stress.

8.3.1 Photosynthetic Compensation

Plants often increase photosynthesis in unaffected tissues to offset damage caused by pest feeding. This compensation mechanism is crucial for maintaining biomass and yield (Li et al. 2019).

8.3.2 Biochemical Adjustments

1. *Antioxidant Enzymes*: Tolerant mungbean varieties upregulate antioxidant enzymes such as superoxide dismutase (SOD) and catalase (CAT) to mitigate oxidative damage caused by pest attack (Kpoviessi et al. 2021).
2. *Osmolyte Accumulation*: Compounds like proline and glycine betaine help maintain cellular stability and turgor, allowing the plant to survive under pest-induced stress (Dhaliwal et al. 2023).

8.3.3 Hormonal Crosstalk

Hormones such as abscisic acid (ABA) and cytokinin are involved in maintaining growth and water balance during pest infestation, contributing to tolerance (War et al. 2020).

8.3.4 Molecular Pathways

Stress-responsive pathways involving heat shock proteins (HSPs) and dehydrins are activated in tolerant plants to protect cellular structures and ensure recovery after pest damage (Li et al. 2019). The insights into antibiosis, antixenosis, and tolerance mechanisms form the foundation for molecular breeding approaches.

9 Molecular Breeding Tools for Pest Resistance in Mungbean

9.1 Marker-Assisted Selection (MAS)

MAS enables the identification and incorporation of pest resistance traits by using molecular markers linked to resistance genes. This approach allows breeders to identify desirable traits in early growth stages, bypassing traditional phenotypic evaluations.

9.1.1 MYMV Resistance

Resistance to mungbean yellow mosaic virus (MYMV) has been a significant focus. Molecular markers linked to the VrMYMIV-1 gene have been used for MAS, leading to the development of resistant varieties (Wu et al. 2022). The VrMYMIV-1 gene has been successfully introgressed into elite varieties through MAS. This approach has reduced breeding cycles and enhanced the precision of selection (Mukhtar and Mohamad 2022).

9.1.2 Bruchid Resistance

Resistance loci linked to bruchid resistance, such as qBru1, have been identified, facilitating the breeding of bruchid-resistant varieties, improving storage resilience (Wu et al. 2022).

9.1.3 Advantages

MAS accelerate the breeding process by reducing the reliance on multi-season field evaluations, thereby saving time and resources. It also enhances selection accuracy, especially for traits with low heritability.

9.1.4 Challenges

Limited availability of molecular markers for some pest resistance traits in mungbean is a constraint. Nonetheless, advances in genome sequencing have mitigated this issue.

9.2 Quantitative Trait Loci (QTL) Mapping

QTL mapping identifies genomic regions associated with complex traits, such as pest resistance, by correlating phenotypic and genotypic data.

9.2.1 MYMV Resistance

Identifying QTLs associated with pest resistance traits, such as trichome density or pod hardness, allows for efficient selection (Mathivathana et al. 2019). QTLs linked to MYMV resistance have been mapped to specific chromosomes in mungbean, aiding in marker development (Mishra et al. 2020). QTLs such as qMYMIV-6-1 have been identified for MYMV resistance in mungbean populations, facilitating marker-assisted backcrossing (Sudha et al. 2022).

9.2.2 Bruchid Resistance

QTLs linked to resistance against *Callosobruchus* spp. have been mapped, enabling fine mapping and functional validation of candidate genes (Wu et al. 2022). For example, researchers in India identified and fine-mapped resistance loci for MYMV, leading to the development of resistant mungbean lines through introgression breeding (Mishra et al. 2020).

9.2.3 Benefits

- Pinpoints genomic regions for breeding programs.
- Accelerates the identification of candidate genes for cloning or gene editing.

9.2.4 Limitations

QTL effects can vary across environments, necessitating multi-location trials for validation.

9.3 Genomic Selection (GS)

Genomic selection (GS) leverages high-density genome-wide markers to predict breeding values and select superior lines without extensive phenotyping. GS uses whole-genome marker information to predict breeding line performance. It allows for simultaneous selection of multiple traits, reducing reliance on phenotypic evaluation (Sodedji et al. 2020). Integrating genomic data with phenotypic selection improves breeding accuracy and accelerates the development of pest-resistant mungbean varieties (War et al. 2018).

9.3.1 Applications in Mungbean

- Resistance to multiple pests, such as bruchids and whiteflies, can be simultaneously targeted using GS models.
- GS enables breeding for pest resistance alongside yield, drought tolerance, and other agronomic traits (Kundu et al. 2022).

9.3.2 Advantages

- Simultaneous selection for polygenic traits.
- Reduces phenotyping costs and time.

9.3.3 Practical Example

In legume breeding programs, GS has been integrated with phenotypic data to predict resistance to complex traits, such as pod borer resistance in chickpea, which can serve as a model for mungbean (Sudha et al. 2022).

9.4 Genetic Engineering

Transgenic approaches introduce pest-resistance genes into mungbean.

9.4.1 Bt Toxins

Genes encoding Cry proteins from *Bacillus thuringiensis* (Bt) have been used to confer resistance against *Marucavitrata* (pod borer) in other legumes, providing a roadmap for mungbean (Purushothaman et al. 2013).

9.4.2 RNA Interference (RNAi)

RNAi-based silencing of critical pest genes has been explored to disrupt pest feeding and reproduction. For instance, silencing gut protease genes in bruchids using RNAi can reduce pest fitness (Nair et al. 2019). RNAi-based silencing of essential pest genes is under development to target aphids and pod borers (Bharathi et al. 2023).

9.4.3 Examples from Other Crops

Transgenic cowpea expressing Bt Cry1Ab protein has shown resistance to pod borers, demonstrating the feasibility of similar approaches in mungbean (Pratap et al. 2021).

9.4.4 Challenges

- Regulatory hurdles and consumer acceptance.
- High costs associated with transformation and biosafety evaluation.

9.5 Genome Editing (CRISPR/Cas9)

CRISPR/Cas9 offers precise editing of pest-susceptibility genes, allowing the development of resistant varieties without introducing foreign DNA.

9.5.1 Applications in MYMV Resistance

CRISPR has been employed to disrupt susceptibility genes, improving resistance durability (Mukhtar and Mohamad 2022).

9.5.2 Host-Pest Interaction Genes

Editing genes involved in pest feeding and oviposition could provide long-term pest resistance solutions. CRISPR/Cas9 technology enables precise modifications of genes regulating pest resistance, enhancing traits such as secondary metabolite biosynthesis (Bhowmik et al. 2021). Modifying genes that regulate pest-host interactions, such as those involved in VOC biosynthesis, can deter pest colonization (Kundu et al. 2022).

9.5.3 Disruption of Susceptibility Genes

Editing genes involved in MYMV entry pathways, such as eIF4E, can enhance resistance (Mukhtar and Mohamad 2022).

9.5.4 Practical Example

Genome editing has been used in tomato to knock out genes facilitating whitefly feeding. Similar strategies can be applied to mungbean (Wisessing et al. 2008).

9.5.5 Advantages

- Precision targeting of resistance traits.
- Avoids issues related to transgenic technologies, enhancing acceptance.

9.5.6 Challenges

Limited genomic resources and editing efficiency in mungbean are barriers to widespread application.

10 Success Stories in Molecular Breeding for Pest Resistance

10.1 Development of MYMV-Resistant Varieties

Varieties such as 'ML 267' and 'Meha', developed through MAS, exhibit high resistance to MYMV and are widely cultivated in India (Mukhtar and Mohamad 2022).

10.2 Bruchid-Resistant Lines

MAS have enabled the development of bruchid-resistant mungbean lines with significantly reduced storage losses.

10.3 Transgenic Approaches for Pod Borers

Although still under trial, Bt mungbean varieties show potential for field-level resistance against pod borers (Kaur et al. 2021).

11 Resistance to Bruchids

11.1 qBru1 Locus

Description: SSR markers linked to the qBru1 locus have enabled breeding of bruchid-resistant mungbean varieties.
Implementation: MAS has helped integrate this locus into commercial lines, reducing seed damage during storage (Choudhary et al. 2024).

11.2 Bruchid Resistance from Wild *Vigna* Species

Gene Source: Resistance loci from *Vigna umbellata* were transferred to mungbean.
Result: MAS facilitated introgression of these loci, resulting in bruchid-resistant mungbean lines with high germination rates post-storage (Rodríguez-Sifuentes et al. 2020).

12 Resistance to Pod Borers (*Maruca vitrata*)

12.1 Cry1Ac Transgenic Screening

Description: Although not yet widely implemented in mungbean, MAS has identified resistance-associated markers in Bt-transgenic legumes such as cowpea.
Future potential: These markers are being tested for MAS-driven resistance in mungbean (Sudha et al. 2022).

12.2 Host Plant Resistance QTLs

Source: Pod borer-resistant traits have been mapped in pigeonpea and yardlong bean.
Application: The markers linked to these QTLs are being transferred to mungbean for improved resistance (Sharma et al. 2008).

13 Resistance to Whiteflies (*Bemisia tabaci*)

13.1 QTL Identification for MYMV Vector Resistance

Description: Whitefly resistance QTLS have been identified and linked with SSR markers.

Examples: MAS efforts have developed whitefly-tolerant mungbean lines to minimize MYMV spread (Nair et al. 2019).

13.2 Integration of Whitefly Resistance from Wild Relatives

Source: *Vigna trilobata*, a wild relative, was used as a donor for whitefly resistance.

Outcome: MAS facilitated introgression into high-yielding mungbean varieties, showing reduced vector transmission rates (Dharajiya and Ravindrababu 2019).

14 Resistance to Aphids (*Aphis craccivora*)

14.1 Antixenosis QTLs for Aphid Resistance

Description: Markers linked to antixenosis traits against aphids have been identified in mungbean.

Implementation: MAS has utilized these markers to incorporate resistance into breeding lines (Sharma and Crouch 2004).

14.2 Breeding Aphid-Resistant Lines

Source: Resistance traits from *Vigna mungo* were mapped and introgressed into mungbean varieties using MAS.

Result: Reduced aphid colonization and better yield stability were observed (Mofokeng and Gerrano 2021).

15 Challenges and Future Directions

Mungbean breeding for pest resistance faces several challenges, including limited genomic resources, which hinder the identification and deployment of resistance genes compared to major crops. The continuous evolution of pests exacerbates the problem, necessitating the stacking of multiple resistance genes to ensure durable protection. Furthermore, transgenic and genome-edited crops face stringent regulatory scrutiny, delaying their adoption. To overcome these challenges, future efforts should focus on developing durable resistance by stacking resistance genes through marker-assisted selection (MAS) and genome editing. Integrated pest management

(IPM) approaches that combine molecular breeding with cultural and biological control measures can offer holistic and sustainable pest management strategies. Additionally, advanced genomics techniques, such as transcriptomics and proteomics, should be leveraged to uncover novel resistance mechanisms and accelerate the development of pest-resistant mungbean varieties.

16 Conclusion

Molecular breeding represents a transformative approach to enhance pest resistance in mungbean, addressing the limitations of traditional methods. By integrating tools like MAS, QTL mapping, genomic selection, and genome editing, researchers can develop high-yielding, pest-resistant mungbean varieties that ensure food security and environmental sustainability. Collaborative efforts between researchers, policymakers, and farmers are essential to harness the full potential of molecular breeding and secure the future of mungbean production in pest-prone regions.

References

Bharathi JK, Anandan R, Benjamin LK, Muneer S, Prakash MAS (2023) Recent trends and advances of RNA interference (RNAi) to improve agricultural crops and enhance their resilience to biotic and abiotic stresses. Plant Physiol Biochem 194:600–618. https://doi.org/10.1016/j.plaphy.2022.11.035

Bhowmik P, Konkin D, Polowick P, Hodgins CL, Subedi M, Xiang D, Yu B, Patterson N, Rajagopalan N, Babic V (2021) CRISPR/Cas9 gene editing in legume crops: opportunities and challenges. Legume Sci 3(3):e96. https://doi.org/10.1002/leg3.96

Bohra A, Jha UC, Kishor PK, Pandey S, Singh NP (2014) Genomics and molecular breeding in lesser explored pulse crops: current trends and future opportunities. Biotechnol Adv 32(8):1410–1428. https://doi.org/10.1016/j.biotechadv.2014.09.001

Cheema HK, Pratap A, Sujayanand G (2017) Breeding for insect resistance in Mungbean and urd bean. In: Breeding insect resistant crops for sustainable agriculture, pp 353–385. https://doi.org/10.1007/978-981-10-6056-4_12

Choudhary KB, Singh D, Jadon K, Solanki R, Kakani R (2024) Biotic stresses in Mungbean: achievements and prospects for genomics-enabled breeding strategies. In: Genomics-aided breeding strategies for biotic stress in grain legumes, pp 135–162. https://doi.org/10.1007/978-981-97-3917-2_5

Dhaliwal SK, Pandey A, Lahkar C, Sheetal, Kaur S (2023) Biotic stress resistance in *Vigna mungo* and *Vigna radiata*: a molecular perspective. In: Diseases in legume crops: next generation breeding approaches for resistant legume crops. Springer, pp 141–173. https://doi.org/10.1007/978-981-99-3358-7_7

Dharajiya DT, Ravindrababu Y (2019) Identification of molecular marker associated with Mungbean yellow mosaic virus resistance in Mungbean [*Vigna radiata* (L.) Wilczek]. Vegetos 32(4):532–539. https://doi.org/10.1007/s42535-019-00063-y

Divekar PA, Rani V, Majumder S, Karkute SG, Molla KA, Pandey KK, Behera TK, Govindharaj GPP (2023) Protease inhibitors: an induced plant defense mechanism against herbivores. J Plant Growth Regul 42(10):6057–6073. https://doi.org/10.1007/s00344-022-10767-2

Dogimont C, Bendahmane A, Chovelon V, Boissot N (2010) Host plant resistance to aphids in cultivated crops: genetic and molecular bases, and interactions with aphid populations. C R Biol 333(6–7):566–573. https://doi.org/10.1016/j.crvi.2010.04.003

Guo T (2024) The role of genomics in advancing pulse crop productivity. Legume Genom Genet 18:15

Kaur R, Choudhury A, Chauhan S, Ghosh A, Tiwari R, Rajam MV (2021) RNA interference and crop protection against biotic stresses. Physiol Mol Biol Plants 27(10):2357–2377. https://doi.org/10.1007/s12298-021-01064-5

Kebede E (2021) Contribution, utilization, and improvement of legumes-driven biological nitrogen fixation in agricultural systems. Front Sustain Food Syst 5:767998. https://doi.org/10.3389/fsufs.2021.767998

Khan MK, Pandey A, Athar T, Hamurcu M, Gezgin S, Sassi G, Kahraman A, von Wettberg EB, Rajput VD, Singh A (2022) Current trends in genetic enhancement of legumes in the genomics era for a sustainable future. In: Advances in legumes for sustainable intensification, pp 533–552. https://doi.org/10.1016/B978-0-323-85797-0.00027-6

Kpoviessi AD, Agbahoungba S, Agoyi EE, Nuwamanya E, Assogbadjo AE, Chougourou DC, Adoukonou-Sagbadja H (2021) Primary and secondary metabolite compounds in cowpea seeds resistant to the cowpea bruchid [*Callosobruchus maculatus* (F.)] in postharvest storage. J Stored Prod Res 93:101858. https://doi.org/10.1016/j.jspr.2021.101858

Kundu A, Ganguli S, Pal A (2022) Genomic designing towards biotic stress resistance in Mungbean and urdbean. In: Genomic designing for biotic stress resistant pulse crops. Springer, pp 381–414. https://doi.org/10.1007/978-3-030-91043-3_8

Lal S (1985) A review of insect pests of Mungbean and their control in India. Int J Pest Manag 31(2):105–114. https://doi.org/10.1080/09670878509370960

Li N, Han X, Feng D, Yuan D, Huang LJ (2019) Signaling crosstalk between salicylic acid and ethylene/jasmonate in plant defense: do we understand what they are whispering? Int J Mol Sci 20(3):671. https://doi.org/10.3390/ijms20030671

Lin YP, Chen HW, Yeh PM, Anand SS, Lin J, Li J, Noble T, Nair R, Schafleitner R, Samsononova M (2023) Demographic history and distinct selection signatures of two domestication genes in Mungbean. Plant Physiol 193(2):1197–1212. https://doi.org/10.1093/plphys/kiad356

Mathivathana MK, Murukarthick J, Karthikeyan A, Jang W, Dhasarathan M, Jagadeeshselvam N, Sudha M, Vanniarajan C, Karthikeyan G, Yang TJ (2019) Detection of QTLs associated with Mungbean yellow mosaic virus (MYMV) resistance using the interspecific cross of *Vigna radiata* × *Vigna umbellata*. J Appl Genet 60:255–268. https://doi.org/10.1007/s13353-019-00506-x

Mishra GP, Dikshit HK, Sv R, Tripathi K, Kumar RR, Aski M, Singh A, Roy A, Priti and Kumari N. (2020) Yellow mosaic disease (YMD) of Mungbean (*Vigna radiata* (L.) Wilczek): current status and management opportunities. Front Plant Sci 11:918. https://doi.org/10.3389/fpls.2020.00918

Mofokeng MA, Gerrano AS (2021) Efforts in breeding cowpea for aphid resistance: a review. Acta Agric Scandinavica, Sect B 71(6):489–497. https://doi.org/10.1080/09064710.2021.1923797

Mukhtar S, Mohamad EAA (2022) Perspectives of molecular marker-assisted breeding in Mungbean (*Vigna radiata* L.) to develop resistance for Mungbean yellow mosaic virus (MYMV) disease. Phytopathogenom Dis Control 1:1–24. https://doi.org/10.22194/pdc/1.1234

Nair RM, Pandey AK, War AR, Hanumantharao B, Shwe T, Alam A, Pratap A, Malik SR, Karimi R, Mbeyagala EK (2019) Biotic and abiotic constraints in Mungbean production progress in genetic improvement. Front Plant Sci 10:1340. https://doi.org/10.3389/fpls.2019.01340

Praharaj S, Maitra S (2020) Importance of legumes in agricultural production systems: an overview. Agro Economist 7(2):69–71

Pratap A, Das A, Kumar S, Gupta S (2021) Current perspectives on introgression breeding in food legumes. Front Plant Sci 11:589189. https://doi.org/10.3389/fpls.2020.589189

Purushothaman S, Nair AS, Renjan B, Karthikeyan K (2013) Efficacy of *Bacillus thuringiensis* against *Maruca vitrata* (Geyer) on cowpea. J Biol Control 27(3):214–216

Rodríguez-Sifuentes L, Marszalek JE, Chuck-Hernández C, Serna-Saldívar SO (2020) Legume protease inhibitors as biopesticides and their defense mechanisms against biotic factors. Int J Mol Sci 21(9):3322. https://doi.org/10.3390/ijms21093322

Sehgal V, Ujagir R (1988) Insect pests and pest management of Mungbean in India. Mungbean:315–328

Sharma H, Crouch J (2004) Molecular marker-assisted selection: a novel approach for host plant resistance to insects in grain legumes. pp 147–174

Sharma H, Varshney R, Gaur P, Gowda C (2008) Potential for using morphological, biochemical, and molecular markers for resistance to insect pests in grain legumes. J Food Legumes 21(4):211–217

Sodedji FAK, Agbahoungba S, Nguetta SPA, Agoyi EE, Ayenan MAT, Sossou SH, Mamadou C, Assogbadjo AE, Kone D (2020) Resistance to legume pod borer (*Maruca vitrata* Fabricius) in cowpea: genetic advances, challenges, and future prospects. J Crop Improv 34(2):238–267. https://doi.org/10.1080/15427528.2019.1680471

Soundararajan R, Chitra N, Geetha S (2013) Host plant resistance to insect pests of grain legumes—a review. Agric Rev 34(3):176–187. https://doi.org/10.5958/j.0976-0741.34.3.002

Sudha M, Karthikeyan A, Madhumitha B, VeeraRanjani R, Mathivathana MK, Dhasarathan M, Murukarthick J, Samu Shihabdeen MN, Eraivan Arutkani Aiyanathan K, Pandiyan M (2022) Dynamic transcriptome profiling of Mungbean genotypes unveils the genes responding to Mungbean yellow mosaic virus infection. Pathogens 11(2):190. https://doi.org/10.3390/pathogens11020190

Varshney RK, Thudi M, Pandey MK, Tardieu F, Ojiewo C, Vadez V, Whitbread AM, Siddique KHM, Nguyen HT, Carberry PS (2018) Accelerating genetic gains in legumes for the development of prosperous smallholder agriculture: integrating genomics, phenotyping, systems modelling and agronomy. J Exp Bot 69(13):3293–3312. https://doi.org/10.1093/jxb/ery088

War AR, Taggar GK, Hussain B, Taggar MS, Nair RM, Sharma HC (2018) Plant defense against herbivory and insect adaptations. AoB Plants 10(4):ply037. https://doi.org/10.1093/aobpla/ply037

War AR, Buhroo AA, Hussain B, Ahmad T, Nair RM, Sharma HC (2020) Plant defense and insect adaptation with reference to secondary metabolites. In: Co-evolution of secondary metabolites, pp 795–822

Wisessing A, Engkagul A, Wongpiyasatid A, Chuwongkomon K (2008) Proteomic and characterization of α-amylase inhibitor from Mungbean (*Vigna radiata*). Agric Nat Resour 42(5):245–250

Wu R, Zhang Q, Lin Y, Chen J, Somta P, Yan Q, Xue C, Liu J, Chen X, Yuan X (2022) Marker-assisted backcross breeding for improving bruchid (*Callosobruchus* spp.) resistance in Mungbean (*Vigna radiata* L.). Agronomy 12(6):1271. https://www.mdpi.com/2073-4395/12/6/1271

Molecular Breeding Approaches for Disease Resistance

Huiying Shi and Yingchao Shen

Abstract

Mungbean crops are severely affected by various biotic and biotic factors that can causes significant yield losses across different countries. Diseases like yellow mosaic disease (YMD), powdery mildew, cercospora leaf spot, dry root rot, fusarium wilt, and halo blight reduce the productivity of the mungbean crop. Conventional breeding techniques are normally used to develop resistant varieties against various diseases. However, due to limitations in mungbean breeding programs, combination of molecular tools along with conventional breeding techniques is highly suitable for development of resistant varieties against various diseases. For YMD, three QTLs associated with MYMIV resistance have been identified. The major QTL, PMR1, was identified to have resistance to powder mildew. Another significant QTL, qCLS, located on linkage group 3, explains approximately 65.5–80.53% of the phenotypic variation in Cercospora leaf spot (CLS) resistance.

Keywords

Mungbean · Powdery mildew · *Fusarium* wilt · Yellow Mosaic Disease (YMD) · Powdery mildew

H. Shi (✉)
Institute of Cereal and Oil Crops, Hebei Academy of Agricultural and Forestry Sciences, Shijiazhuang, Hebei, China

Y. Shen
Hebei Key Laboratory of Crop Genetics and Breeding, Shijiazhuang, Hebei, China

S. Paramasivam et al. (eds.), *Mungbean*,
https://doi.org/10.1007/978-981-95-7637-1_5

1 Introduction

Mungbean is susceptible to be infected by several viral, fungal, and bacterial diseases (Noble et al. 2019). Yellow mosaic disease (YMD) is the most serious viral disease that affects mungbean cultivation. Among the fungal diseases, the commonly occurring diseases are powdery mildew (PM), cercospora leaf spot (CLS), dry root rot, and fusarium wilt (FW). Halo blight (*Pseudomonas savastanoi* pv. *phaseolicola*) is an important bacterial disease in mungbean (Fig. 1, Table 1). The development of resistant mungbean varieties against various diseases heavily depends on the availability and identification of resistant germplasm for specific traits. Reliable resistant germplasm forms the crucial foundation for breeding programs aimed at developing disease-resistant mungbean varieties (Mogali and Hegde 2020). By utilizing these resistant sources, breeders can incorporate resistance traits into high-yielding, locally adapted cultivars through conventional breeding techniques. Molecular breeding approaches have revolutionized the development of disease-resistant mungbean varieties by enabling precise selection of desirable traits at the DNA level (Choudhary et al. 2024). A combination of conventional and molecular breeding approaches is highly helpful to the plant breeders of the mungbean varieties with durable disease resistance, ensuring crop stability and productivity.

Fig. 1 (**a**–**f**) Symptoms of major diseases in mungbean. (**a**) Mungbean yellow mosaic disease, (**b**) cercospora leaf spot, (**c**) powdery mildew, (**d**) dry root rot, (**e**) fusarium wilt, (**f**) halo blight

Table 1 Summary of major mungbean diseases, their causal agents, and optimal weather conditions for disease development

Diseases	Causal agents	Optimal weather conditions	Losses	References
Yellow mosaic disease	Mungbean yellow mosaic virus	Humid weather	10–100%	Dubey and Singh (2010)
Cercospora leaf spot	*Cercospora canescens*	Humid weather	Up to 50%	Dubey and Singh (2010)
Powdery mildew	*Erysiphae polygoni*	Humid weather	9–50%	Pandey et al. (2021)
Dry root rot	*Macrophomina phaseolina*	Warmer condition and water stress	10–44%	Choudhary et al. (2011)
Fusarium wilt	*Fusarium oxysporum*	High temperatures and warm moist soil	30–100%	Sun et al. (2019)
Halo blight disease	*Pseudomonas savastanoi* pv. *Phaseolicola*	Cool, wet, and windy climate	30–100%	Rana et al. (2016)

1.1 Yellow Mosaic Disease

Yellow mosaic disease (YMD) is a primary viral disease that reduces mungbean production in South Asia, especially in India and Maynar. The symptoms caused by YMD include leaf yellowing, chlorosis preceded by necrosis, fewer flowers and pods, and abnormal seeds and stunting plants. The disease transmitted by the whitefly *Bemisia tabaci*, causes 10–100% yield reduction (Mishra et al. 2020).

Pathogen of YMD totally is known as yellow mosaic viruses (YMVs) (Table 1). Four different YMVs have been found, i.e., mungbean yellow mosaic virus (MYMV), mungbean yellow mosaic India virus (MYMIV), dolichos yellow mosaic virus (DoYMV), and horse gram yellow mosaic virus (HgYMV). MYMV and MYMIV are mainly pathogens of YMD (Mishra et al. 2020).

Due to the diversity of pathogen strains and plant genotypes, multiple types of YMD resistance in mungbean have been reported, including a single dominant gene or a single recessive gene, or two recessive genes or complementary recessive genes (Shukla and Pandya 1985; Malik et al. 1988; Dhole and Reddy 2012; Alam et al. 2014a; Singh et al. 2020).

The pathogen MYMV is mainly distributed in western and southern India, Thailand, and Indonesia. Selvi et al. (2006) developed OPS 790, a RAPD marker associated with MYMV resistance in ML 267. Dhole and Reddy (2013) identified an SCAR marker, OPB-$07_{600,}$ in resistant material TM-99-37. Yen et al. (2016) reported several MYMV-linked SNPs on the Mungbean chromosomes 2, 5, 7, 9, and 10. CM9 and CM815, two SCAR markers located in chromosome 3, was found to be linked with MYMV resistance gene (Sai et al. 2017). Mathivathana et al. (2019) reported qMYMV4_1 on chromosome 4, a stable and major QTL for resistance to MYMV, using an RIL population of an interspecific cross between Mungbean-VRM (Gg) 1 and Ricebean-TNAU RED. This QTL contains 83 annotated genes of which 18 are considered as candidate genes. A transcriptome analysis was conducted between VGGRU-1 and VRM (Gg) 1, and *Vradi04g06840* encoding small GTP-binding protein was of interest during plant disease resistance (Sudha et al. 2022).

Chen et al. (2013) identified three QTLs associated with MYMIV resistance. Kitsanachandee et al. (2013) developed an RIL population and identified five QTLs for MYMIV resistance—three of the QTLs for India and two for Pakistan, of which qYMIV1 (India) and qYMIV4 (Pakistan) appeared to have the same locus and to be the same with the major QTL of Chen et al. (2013). Alam et al. (2014b) identified two QTLs, qMYMIV2 on linkage group 2 and qMYMIV7 on linkage group 7. The qMYMIV2 appeared to be the same as that identified by Kitsanachandee et al. (2013) and Chen et al. (2013). Singh et al. (2020) detected 15 microsatellite markers associated with MYMIV resistance, of which 3 microsatellites, CEDG211, CEDG121, and DMB-SSR008, accounted for 11–14% phenotypic variation. Rohilla et al. (2022) identified 10 QTLs associated with YMD resistance using the mixed linear model (MLM) method.

Developing mungbean varieties with resistance to the causal virus is a challenge. Many studies evaluated mungbean germplasms that are resistant to MYMV, but few of the accessions are resistant to YMD, such as ML1628, NM-92, NM-94, EC396424, EC396425, EC718740-EC718745 (Table 3). Five genotypes namely AVMU 1698, AVMU 1699, AVMU 16100, AVMU 16101, and KPS 2 were identified as resistant to YMD (Basavaraj et al. 2019). Sudha et al. (2013) identified that mungbean genotype VGGRU-1 is resistant to MYMV, while VRM (Gg) 1 is susceptible (Table 3).

The YMD resistance is generally evaluated by the symptoms using a commonly accepted disease scoring scale (Khattak et al. 2008; Iqbal et al. 2011; Panduranga et al. 2011). Most of the reports about the identification of YMD resistance resources in mungbean are based on field screening. Vairam et al. (2016) have identified five mutants (viz., M5, M18, M26, M70, and M71), which in M_3 generation showed YMD resistance. Akhtar et al. (2009) identified 3 out of 146 recombinant in-bred lines (RILs) developed in Thailand as tolerant using field screening. Akhtar et al. (2011) reported 35 mungbean genotypes among 162 genotypes moderately resistant to MYMIV disease under natural infection in field condition. Gupta and Mishra (2014) reported 54 mungbean genotypes resistant to MYMV disease under natural infection.

1.2 Powdery Mildew

Powdery mildew (PM), caused by the fungus *Erysiphe polygoni* D.C., is a major fungal disease in southeast Asia (Table 1). The disease is especially prevalent in the cool-dry season and usually cause 20–60% yield reduction and even complete loss at the seedling and early stage (Aumentado et al. 2023). The symptoms of PM are white powdery coating on leaves, stems, and pods.

Reddy et al. reported two dominant genes of PM resistance, named *Pm-1* and *Pm-2*, from the resistant line RUM, and a single dominant gene *Pm-3* from resistant variety Mulmarada (Reddy et al. 1994; Reddy 2009). Chaitieng et al. (2002) used RFLP and amplified fragment length polymorphism (AFLP) markers and identified one major and one minor QTL, PMR1, and PMR2, controlling resistance to PM in "VC1210A". The major QTL, PMR1, accounted for 64.9% of the variation in resistance to PM. Humphry et al. (2003) identified a major locus conferring PM resistance from the highly resistant line ATF360; this locus accounted for 86% of the variation. Zhang et al. (2008) developed three SSR markers and two STS markers, using the RFLP marker VrCS65 of Humphry et al. (2003). Kasettranan et al. (2009) identified two QTLs conferring PM resistance, *qPMR1*and *qPMR2*, from the resistant line VC6468–11-1a. The two QTLs accounted for 20.10%and 57.81% variation for PM resistance, respectively. The qPMR2 is possibly the same as found by Humphry et al. (2003). Chankaew et al. (2013) identified one major QTL (qPMV4718-3, on linkage group 9) and two minor QTLs (qPMV4718-1 and qPMV4718-2, both on linkage group 4) from resistant accession V4718, and two major QTLs (qPMRUM5-2 on linkage group 6 and qPMRUM5-3 on linkage group 9) and one mionrQTL (qPMRUM5-1 on linkage group 4) from resistant accession RUM5. Among them, the QTLs on linkage group 9 (qPMV4718-3 and qPMRUM5-3) and a QTL (qPMRUM5-2) on linkage group 6 are the same as, or closely linked to qPMR2 and qPMR1 reported by Kasettranan et al. (2010), respectively. The locus qPMRUM5-3 and qPMRUM5AQ6-2 are possibly the same as the Pm-1 and Pm-2 loci identified by Reddy et al. (1994). Poolsawat et al. (2017) used resistant line V4718 and mapped a major QTL, qPMC72V18–1, and explained 92.4%of phenotype variation, flanked by I42PL229 and I85420 markers. Yundaeng et al. (2020) revealed VrMLO12, which is located in the qPMRUM5–3 region and associated with PM susceptibility in mungbean. Tantasawat et al. (2022) developed molecular marker I41tP379 for mungbean breeding. Waengwan et al. (2024) conducted fine mapping for qPMRUM5-2, located a 135.0 kb region on chromosome 8 containing nine predicted genes of which five were NBS-LRR-type genes *Recognition of Peronospora parasitica 13-like protein* (*RPP13L*).

The development of PM-resistant cultivars is therefore a major goal in mungbean breeding programs. Hegde (1999) reported that DHMC 9601, DHMC 9602, DHMC 9603, and DHMC 9604 were highly resistant to PM. BPMR 1 and BPMR 115 were reported as resistant to powdery mildew (Singh and Gurha 2005). BPMR-145, TARM-18, Phule M-2003-3, Phule M-2002–13, Phule M-2002-17, Phule M-2001-3, and Phule M-2001-5 exhibit combined resistance against powdery mildew and yellow mosaic virus (Mandhare and Suryawanshi 2008).

1.3 Cercospora Leaf Spot

Cercospora leaf spot (CLS) caused by the fungus *Cercospora cruenta* and *C. canescens* is an important disease of mungbean (Table 1), which resulted in 50% seed yield reduction (Dubey and Singh 2010). CLS disease often appears under high humidity conditions. The characteristic symptoms are small leaf spots (1–5 mm) with brown to grayish centers and reddish border.

Many mungbean germplasms have been reported to possess resistance to CLS, such as Pusa 105, PDM 15, PDM 2, PDM 113, PDM 14 (Mishra et al. 1988). Hassan et al. (2017) found that VC-6153-B-20 genotype was identified as a CLS tolerant variety. 10-43 (NM89) and 10-12 (NM88) in particular are not only high yielding and early maturing genotypes but are also resistant to CLS diseases (Ali et al. 1997).

CLS resistance of mungbean was governed by a dominant gene (Thakur et al. 1977). Mishra et al. (1988) detected a 3:1 ratio in 14 crosses of involving resistant × susceptible parents, suggesting the CLS resistance was governed by a single recessive gene, which is similar to that identified by Akhtar (2023). Researchers identified that V4718 showed high and stable resistance to CLS, and the resistance in V4718 is controlled by a single dominant gene (Hartman et al. 1993). Chankaew et al. (2011) identified a major QTL, qCLS, on linkage group 3 and accounted for 65.5–80.53% variation of CLS resistance. Yundaeng et al. (2021) fine-mapped qCLS to a 13 kb region on chromosome 6, which contains only one annotated gene (LOC106765332) encoding a TATA-binding-protein-associated factor 5 (TAF5). Yadav et al. (2014) screened 65 mungbean genotypes, among which LGG-460 was found to be highly resistant to CLS, while GM-02-08, GM-02-13, and GM-03-03 were categorized as resistant.

1.4 Dry Root Rot Disease

Dry root rot (DRR), caused by *Macrophomina phaseolina*, is an emerging disease of mungbean (Table 1), which results in wilting in advanced stages and ultimately causes the death of plants (Choudhary et al. 2011). DRR causes huge losses in the production by reducing plant population at both seedling and adult stages (Khan et al. 2016). In addition, the disease results in large losses in the premium sprout market (Fuhlbohm et al. 2013). Iqbal et al. (2011) and Kumar et al. (2017) detected *Macrophomina phaseolina* isolates from Pakistan and India, respectively. The pathogenicity of isolates showed clear diversity. In recent years, researchers have made many efforts by screening mungbean accessions for DRR resistance (Table 2).

Table 2 Nature of inheritance of the major diseases of mungbean

Diseases	Number of genes and nature of inheritance	References
Yellow mosaic disease	A single dominant gene or a single recessive gene, or two recessive genes or complementary recessive genes	(Shukla and Pandya (1985), Malik et al. (1988), Dhole and Reddy (2012), Alam et al. (2014a), Singh et al. (2020)
Powdery mildew	Two dominant genes	Reddy et al. (1994)
	Single dominant gene	Reddy (2009)
Cercospora leaf spot	Dominant gene	Thakur et al. (1977)

For example, Pandey et al. (2021) revealed the stability of accessions VI001509AG, VI001244AG, and VI001400AG for DRR resistance. Avanija et al. (2023) screened 47 mungbean genotypes and identified 6 genotypes, viz., MGG-529, MG-549, WGG-42, MG-505, Pusa 9072, and WGG-25, that showed a resistant response to dry root rot. Pandey et al. (2021) identified three root rot-resistant genotypes showing lower disease score, namely IPM99–125, EC693368, and EC693369.

1.5 *Fusarium* Wilt

Fusarium wilt (FW), a fungal disease, is a significant threat to mungbean production, impacting more than 30% yield reduction, even complete loss (Table 1). The disease is mainly transmitted throuhg soil, and its chlamydospores can survive in soil for a long time. and its chlamydospores can survive in soil for a long time. Chlamydospores infect and colonize the root, then gain access to the vascular system, leading to systemic infection, including leaf chlorosis, wilting and necrosis, vascular discoloration and stunting of the whole plant.

FW was a minor disease of mungbean initially. In recent years, the severity and incidence of the disease appeared to be rising in mungbean producing area, including Australia, India, and China (Kelly 2018; Sun et al. 2019; Vani et al. 2019). Pathogen of FW isolated from Australia (Kelly 2018) and China (Sun et al. 2019) was identified as *Fusarium oxysporum*, which has a broad range and has been divided into various *formae speciales*. However, pathogen from Pakistan was identified as *Fusarium nanum* (Kamran et al. 2024).

In mungbean, few studies have been executed to understand the molecular mechanism involved in the response to fusarium wilt infection. Sun et al. (2020) constructed an evaluation method of FW resistance in mungbean. By this method, 11 mungbean cultivars were selected as hosts and the isolates were divided into 12 pathotypes. Chang et al. (2021) selected 27 candidate genes of FW resistance by transcriptome analysis using Zheng8-4 (highly resistant) and Zheng8-20 (highly susceptible), includingresistance proteins, encoding protein kinases, transcription factors and protein with oxidoreductase activity.

1.6 Halo Blight

Halo blight is an important seed-borne bacterial disease in mungbean. The disease favors cool, wet, and windy climate and usually occurs during and after flowering stage. Symptoms of halo blight appears as small, dark, circular spots initially, which then develop into necrotic lesions surrounded by yellow margins. Eventiually, the lesions coalesce and develop into large necrotic areas scattered on both sides of the infected leaf (Abdullah and Douglas 2021). The disease is mainly transmitted by infected seed and other spreading routes including rain splash, plant-to-plant contact, and through irrigation water and filed workers (Arnold et al. 2011). Halo blight distributed in Pakistan (Akhtar 1988), China (Sun et al. 2019), Australia (Noble et al. 2019) can cause significant yield losses that vary from 30% to 100% (Rana et al. 2016).

The causative agent of halo blight is *Pseudomonas savastanoi* pv. *phaseolicola.* Fllowing revisions to the genomospecies classification, the pathogen has been successicely named *Phytomonas medicaginis* variant *phaseolicola*, *Pseudomonas medicaginis* variant *phaseolicola*, and *Pseudomonas syringae* pathovar. *Phaseolicola.* The *P. syringiae* pv. *phaseolicola* have a wide range of hosts, including many legume plants such as common bean, pea, pigeon pea, adzuki bean, mungbean, tick clover, and lima bean (Noble et al. 2019). Taylor et al. (1996) used 8 different cultivars and categorized 175 isolates of *P. syringiae phaseolicola* into 9 races. Among them, mungbean is sensitive to race 1 and 7. Noble et al. (2020) detected 352 isolates from Australia. Isolates revealed five distinctive pathotypes, and pathotype 2 was the most prevalent at 84% and was found to be highly pathogenic toward all four tested mungbean genotypes. Genetic testing for 205 isolates divided the pathogen into two genetic lineages.

2 Conclusion

Molecular breeding approaches have emerged as powerful tools for enhancing disease resistance in mungbean, offering precise, efficient, and rapid development of resistant varieties (Mogali & Hegde 2020). Techniques such as marker-assisted selection (MAS), genomic selection, and gene editing enable the identification and incorporation of resistance genes against major pathogens like mungbean yellow mosaic virus, powdery mildew, and rust. These approaches facilitate the pyramiding of multiple resistance genes, thereby increasing durability and broad-spectrum resistance. Overall, molecular breeding holds significant promise for sustainable mungbean production by reducing yield losses caused by diseases, ensuring food security, and minimizing reliance on chemical control methods. Continued research and integration of molecular tools will further accelerate the development of resilient mungbean cultivars suited to diverse agro-ecological conditions (Fig. 1 and Table 3).

Table 3 Sources of resistance to major diseases in mungbean

Trait	Resistant accession ID	References
YMD	NM-92	Ali et al. (1997)
	PDM-139 (Samrat), IPM-02-03, Pusa0672, HUM16, MB-57, MB-58, PantMung 2	Datta et al. (2012), Paul et al. (2013), Kumar and Kumar (2014), Subedi et al. (2016)
	VGGRU-1	Sudha et al. (2013, 2022)
	EC398897, TM-11-07, TM-11-34, PDM-139, IPM-02-03, IPM-02-14, Pusa0672, Pusa0871, CO-7, MH-521	Mohan et al. (2014)
	ML1628, NM-94	Nair et al. (2017)
	IPM0203, KM2241, PDM139, Pusa0672, HUM16, ML1464, TARM-1	Bhanu et al. (2017)
	SML1815, MH421	Mahalingam et al. (2018)
	AVMU 1698, AVMU 1699, AVMU 16100, AVMU 16101, KPS 2	Basavaraj et al. (2019)
	EC396424, EC396425, EC718740, EC718741, EC718742, EC718743, EC718744, EC718745	Nair et al. (2017)
Cercospora leaf spot	10-43(NM89), 10-12(NM88) (Pakistan)	Ali et al. (1997)
	ML-231, ML-5, ML-267, ML-337	Marappa (2008)
	LGG-460	Yadav et al. (2014)
	VC-6153-B-20	Hassan et al. (2017)
	AVMU-1687, AVMU-1683, AVMU-1698, etc	Sahoo et al. (2022)
Powdery mildew	DHMC 9601, DHMC 9602, DHMC 9603, and DHMC 9604	Hegde (1999)
	BPMR 1 and BPMR 115	Singh and Gurha (2005)
	BPMR-145, TARM-18, Phule M-2003-3, Phule M-2002-13, Phule M-2002-17, Phule M-2001-3, and Phule M-2001-5	Mandhare and Suryawanshi (2008)
	V4718	Chankaew et al. (2011)
Dry root rot	NCM 252-10, 40536 (highly resistant) 40504, NCM 257-5, 40457, NCM 251-4, 6368-64-72 (resistant)	Khan and Shuaib (2007)
	MSJ-118, KM 4-44, KM 4-59	Choudhary et al. (2011)
	Azri 2006, NM 2006, and AUM 9	Haseeb et al. (2013)
	MNUYT-317, NM-2011	Khan et al. (2016)
	IPM99-125, EC693368, EC693369 VI001509AG, VI001244AG, VI001400AG	Pandey et al. (2021)
	MGG-529, MG-549, WGG-42, MG-505, Pusa 9072, WGG-25	Avanija et al. (2023)

References

Abdullah AS, Douglas C (2021) Halo blight of Mungbean in Australia. Plants 1(1):3–7. https://doi.org/10.3390/crops1010002

Akhtar MA (1988) Pseudomonas syringaepv. phaseolicola on Mungbean. Pak J Agric Res 9:424–426

Akhtar MA (2023) Genetics of cercospora leaf spot resistance in Mungbean. SABRAO J Breed Genet 55(4):1109–1122. https://doi.org/10.54910/sabrao2023.55.4.8

Akhtar KP, Kitsanachandee R, Srinives P, Abbas G, Asghar MJ, Shah TM, Atta BM (2009) Field evaluation of Mungbean recombinant inbred lines against Mungbean yellow mosaic disease using new disease scalein Thailand. Plant Pathol J 25(4):422–428. https://doi.org/10.5423/PPJ.2009.25.4.422

Akhtar KP, Sarwar G, Abbas G, Asghar MJ, Sarwar N, Shah TM (2011) Screening of Mungbean germplasm against Mungbean yellow mosaic India virus and its vector Bemisiatabaci. Crop Prot 30:1202–1209. https://doi.org/10.1016/j.cropro.2011.05.012

Alam AKM, Somta P, Srinives P (2014a) Generation mean and path analyses of reaction to Mungbean yellow mosaic virus (MYMV) and yield-related traits in Mungbean (*Vigna radiata* (L.) Wilczek). SABRAO J Breed Genet 46(1):150–159

Alam AKMM, Somta P, Srinives P (2014b) Identification and confirmation of quantitative trait loci controlling resistance to Mungbeanyellow mosaic disease in Mungbean [*Vigna radiata* (L.) Wilczek]. Mol Breed 34(3):1497–1506. https://doi.org/10.1007/s11032-014-0133-0

Ali MM, Sabir IA, Ahmed HM (1997) The Mungbean green revolution in Pakistan. Asian Vegetable Research and Development Centre, Shanhua 24:66

Arnold DL, Lovell HC, Jackson RW, Mansfield JW (2011) *Pseudomonas syringae*pv. *phaseolicola*: from 'has bean' to supermodel. Mol Plant Pathol 12(7):617–627. https://doi.org/10.1111/j.1364-3703.2010.00697.x

Aumentado HDR, Jonathan JGG, Balendres MAO (2023) Powdery mildew in Southeast Asia: host range, impact, and sources of resistance. Technol Agron 3(1). https://doi.org/10.48130/TIA-2023-0009

Avanija M, Padmaja G, Vidyasagar B, Hari Y, Devi GU, Hemantha J (2023) Screening of genotypes against dry root rot of Mungbean (*Vigna radiata* L.) R. Wilczek. Pharma Innov J 12:805–809

Basavaraj S, Padmaja AS, Nagaraju N, Ramesh S (2019) Identification of stable sources of resistance to Mungbean yellow mosaic virus (MYMV) disease in Mungbean [*Vigna radiata* (L.) Wilczek]. Plant Genet Resour 17:362–370. https://doi.org/10.1017/S1479262119000121

Bhanu AN, Singh MN, Srivastava K (2017) Screening Mungbean [*Vigna radiata* (L.) Wilczek] genotypes for Mungbean yellow mosaic virus resistance under natural condition. Adv Plants Agric Res 7(6):00276

Chaitieng B, Kaga A, Han OK, Wang XW, Wongkaew S, Laosuwan P, Tomooka N, Vaughan DA (2002) Mapping a new source of resistance to powdery mildew in Mungbean. Plant Breed 121(6):521–525. https://doi.org/10.1046/j.1439-0523.2002.00751.x

Chang Y, Sun F, Sun S, Wang L, Wu J, Zhu Z (2021) Transcriptome analysis of resistance to Fusarium Wilt in Mungbean (*Vigna radiata* L.). Front Plant Sci 12:679629. https://doi.org/10.3389/fpls.2021.679629

Chankaew S, Somta P, Sorajjapinun W, Srinives P (2011) Quantitative trait loci mapping of Cercospora leaf spot resistance in Mungbean, *Vigna radiata* (L.) Wilczek. Mol Breed 28:255–264. https://doi.org/10.1007/s11032-010-9478-1

Chankaew S, Somta P, Isemura T, Tomooka N, Kaga A, Vaughan DA, Srinives P (2013) Quantitative trait locus mapping reveals conservation of major and minor loci for powdery mildew resistance in four sources of resistance in Mungbean [*Vigna radiata* (L.) Wilczek]. Mol Breed 32:121–130. https://doi.org/10.1007/s11032-013-9856-6

Chen HM, Ku HM, Schafleitner R, Bains TS, George Kuo C, Liu CA, Nair RM (2013) The major quantitative trait locus for Mungbean yellow mosaic Indian virus resistance is tightly linked in repulsion phase to the major bruchid resistance locus in a cross between Mungbean [*Vigna radiata* (L.) Wilczek] and its wild relative *Vigna radiata* ssp. sublobata. Euphytica 192(2):205–216. https://doi.org/10.1007/s10681-012-0831-9

Choudhary S, Choudhary AK, Sharma OP (2011) Screening of Mungbean (*Vigna radiata*) accessions to identify sources of resistance to DRR. J Food Legum 24:117–119

Choudhary KB, Singh D, Jadon KS, Solanki RK, Kakani RK (2024) Biotic stresses in mung bean: achievements and prospects for genomics-enabled breeding strategies. Genomics-aided breeding strategies for biotic stress in grain legumes 14:135–162

Datta S, Gangwar S, Kumar S, Gupta S, Rai R, Kaashyap M, Singh P, Chaturvedi SK, Singh BB, Nadarajan N (2012) Genetic diversity in selected Indian Mungbean [*Vigna radiata* (L.) Wilczek] cultivars using RAPD markers. Am J Plant Sci 3:1085–1091. https://doi.org/10.4236/ajps.2012.38130

Dhole VJ, Reddy KS (2012) Genetic analysis of resistance to Mungbean yellow mosaic virus in Mungbean (*Vigna radiata*). Plant Breed 131(3). https://doi.org/10.1111/j.1439-0523.2012.01964.x

Dhole VJ, Reddy KS (2013) Development of a SCAR marker linked with a MYMV resistance gene in Mungbean (*Vigna radiata* L. Wilczek). Plant Breed 132:127–132. https://doi.org/10.1111/pbr.12006

Dubey SC, Singh B (2010) Seed treatment and foliar application of insecticides and fungicides for management of cercospora leaf spots and yellow mosaic of Mungbean (*Vigna radiata*). Int J Pest Manag 56:309–314. https://doi.org/10.1080/09670874.2010.496875

Fuhlbohm J, Ryley MJ, Aitken EAB (2013) Infection of Mungbean seed by *Macrophominaphaseolina* is more likely to result from localized pod infection than from systemic plant infection. Plant Pathol 62:1271–1284. https://doi.org/10.1111/ppa.12047

Gupta O, Mishra M (2014) Field resistance in Mungbean and urdbean genotypes against yellow mosaic disease. J Food Legum 27:80–81

Hartman GL, Wang TC, Kim D (1993) Field evaluation of Mungbeans for resistance to Cercospora leaf spot and powdery mildew. Int J Pest Manag 39:418–421. https://doi.org/10.1080/09670879309371833

Haseeb HA, Sahi ST, Safdar Ali SA, Muhammad Fiaz MF (2013) Response of different Mungbean varieties against *Macrophominaphaseolina* (Tassi) Goid and in vitro studies of plant extracts against pathogen. Pak J Phytopathol 25:78–83

Hassan R, Haque AH, Datta J, Nayem J, Diya FZ (2017) Mungbean germplasms resistance to cercospora leaf spot disease and its management. J Sylhet Agric Univ 4(1):43–48

Hegde VS (1999) Evaluation of Mungbean genotypes for resistance to powdery mildew. Indian J Plant Genet Resourc 12(03):408–409

Humphry ME, Magner T, McIntyre CL, Aitken EA, Liu CJ (2003) Identification of a major locus conferring resistance to powdery mildew (*Erysiphepolygoni* DC) in Mungbean (*Vigna radiata* L. Wilczek) by QTL analysis. Genome 46(5):738–744. https://doi.org/10.1139/g03-057

Iqbal U, Iqbal SM, Afzal R, Jamal A, Farooq MA, Zahid A (2011) Screening of Mungbean germplasm against Mungbean yellow mosaic virus (MYMV) under field conditions. Pak J Phytopathol 23(1):48–51

Kamran M, Ullah I, Ehetisham ul Haq M, Shahbaz MU, Abbas H, Asghar MJ, Shahid M, Akhtar KP, Abbas H (2024) First report of wilt caused by *fusarium nanum* (FIESC 25) on Mungbean(*Vignaradiata*) in Pakistan. Plant Dis 108(7):2237. https://doi.org/10.1094/PDIS-04-24-0751-PDN

Kasettranan W, Somta P, Srinives P (2009) Genetics of the resistance to powdery mildew disease in Mungbean (*Vigna radiata* (L.) Wilczek). J Crop Sci Biotechnol 12(1):37–42. https://doi.org/10.1007/s12892-008-0074-4

Kasettranan W, Somta P, Srinives P (2010) Mapping of quantitative trait loci controlling powdery mildew resistance in Mungbean (*Vigna radiata* (L.) Wilczek). J Crop Sci Biotechnol 13(3):155–161. https://doi.org/10.1007/s12892-010-0052-z

Kelly L (2018) Fusarium species associated with grain sorghum and Mungbean in Queensland. MPhil Thesis, the University of Queensland. http://espace.library.uq.edu.au/view/UQ:716902

Khan KSH, Shuaib M (2007) Identification of sources of resistant in Mungbean (*Vigna radiata* L.) against charcoal rot *Macrophomina phaseolina* (Tassi) Goid. 8th African Crop Science Society Conference 8:2101–2102

Khan KA, Shoaib A, Akhtar S (2016) Response of *Vigna radiata* (L.) Wilczek genotypes to charcoal rot disease. Mycopath 14:1–7

Khattak GSS, Iqbal S, Shah SA (2008) Breeding high yielding and disease resistant Mungbean (*Vigna radiata* (L.) Wilczek) genotypes. Pak J Bot 40:1411–1417

Kitsanachandee R, Somta P, Chatchawankanphanich O, Akhtar KP, Shah TM, Nair RM, Bains TS, Sirari A, Kaur L, Srinives P (2013) Detection of quantitative trait loci for Mungbean yellow mosaic India virus (MYMIV) resistance in Mungbean (*Vigna radiata* (L.) Wilczek) in India and Pakistan. Breed Sci 63(4):367–373. https://doi.org/10.1270/jsbbs.63.367

Kumar S, Kumar R (2014) Genetic improvement of Mungbean for yield, nutrition and resistance to stresses-a review. Int J Trop Agric 14.32(3-4):683–687. http://serialsjournals.com/archives.php?journals_id=56

Kumar P, Gaur VK, Meena AK (2017) Screening of different Macrophomina phaseolina isolates on susceptible (RMG-62) variety of Mungbean. Int J Pure Appl Biosci 5(6):698–702

Mahalingam A, Satya VK, Manivannan N, Narayanan SL, Sathya P (2018) Inheritance of Mungbean yellow mosaic virus disease resistance in greengram [*Vigna radiata* (L.) Wilczek]. Int J Curr Microbiol App Sci 7(01):880–885

Malik IA, Ali Y, Saleem M (1988) Incorporation of tolerance to Mungbean yellow mosaic virus-from local germplasm to exotic large seeded Mungbean. In S Shanmungasundrum (ed) Mungbean: Proceedings of the Second International Symposium on Mungbean (pp 297–307). Bangkok: AVRDC. 16–20 November 1987

Mandhare VK, Suryawanshi AV (2008) Dual resistance against powdery mildew and yellow mosaic virus in greengram. Agric Sci Dig 28(1):39–41

Marappa N (2008) Screening of Mungbean genotypes andits wild relatives for resistant sources to Cercosporaleaf spot disease. Asian J Bio Sci 3(2):324–326

Mathivathana MK, Murukarthick J, Karthikeyan A, Jang W, Dhasarathan M, Jagadeeshselvam N, Sudha M, Vanniarajan C, Karthikeyan G, Yang TJ, Raveendran M (2019) Detection of QTLs associated with Mungbean yellow mosaic virus (MYMV) resistance using the interspecific cross of *Vigna radiata*×*Vigna umbellata*. J Appl Genet 60(3–4):255–268. https://doi.org/10.1007/s13353-019-00506-x

Mishra SP, Asthana AN, Yadav L (1988) Inheritance of *Cercospora*leafspot resistance in Mungbean, *Vigna radiata* (L.) Wilczek. Plant Breed 100(3):228–229. https://doi.org/10.1111/j.1439-0523.1988.tb00245.x

Mishra GP, Dikshit HK, Sv R, Tripathi K, Kumar RR, Aski M, Singh A, Roy A, Priti, Kumari N, Dasgupta U (2020) Yellow mosaic disease (YMD) of Mungbean (*Vigna radiata* (L.) Wilczek): current status and management opportunities. Front Plant Sci 11:918. https://doi.org/10.3389/fpls.2020.00918

Mogali SC, Hegde GM (2020) Recent advances in Mungbean breeding: a perspective. In: Accelerated plant breeding, volume 3: food legumes, pp 235–282. https://doi.org/10.1007/978-3-030-47306-8_9

Mohan S, Sheeba A, Murugan E, Ibrahim SM (2014) Screening of Mungbean germplasm for resistance to Mungbean yellow mosaic virus under natural condition. Indian J Sci Technol 7(7):891–896

Nair RM, Götz M, Winter S, Giri RR, Boddepalli VN, Sirari A, Bains TS, Taggar GK, Dikshit HK, Aski M, Boopathi M (2017) Identification of Mungbean lines with tolerance or resistance to yellow mosaic in fields in India where different begomovirus species and different Bemisia tabaci cryptic species predominate. Eur J Plant Pathol 149:349–365. https://doi.org/10.1007/s10658-017-1187-8

Noble TJ, Young AJ, Douglas CA, Williams B, Mundree S (2019) Diagnosis and management of halo blight in Australian Mungbeans: a review. Crop Pasture Sci 70. https://doi.org/10.1071/CP18541

Noble TJ, Young AJ, Kelly LA, Barrerro RA, Douglas CA, Long H, Williams B, Mundree S (2020) Characterisation of the *Pseudomonas savastanoi*pv. *phaseolicola* population found in Eastern Australia associated with halo blight disease in *Vigna radiata*. Australas Plant Pathol 49(43). https://doi.org/10.1007/s13313-020-00722-8

Pandey AK, Yee M, Win MM, Lwin HM, Adapala G, Rathore A, Sheu ZM, Nair RM (2021) Identification of new sources of resistance to dry root rot caused by *Macrophominaphaseolina* isolates from India and Myanmar in a Mungbean mini-core collection. Crop Prot 143:105569. https://doi.org/10.1016/j.cropro.2021.105569

Panduranga GS, Vijayalakshmi K, Reddy KL, Rajashekara H (2011) Evaluation of Mungbean germplasm for resistance against whitefly (*Bemisiatabaci Genn.*) and Mungbean yellow mosaic virus (MYMV) disease. Indian J Entomol 73(4):338–342

Paul PC, Biswas MK, Mandal DI, Pal PI (2013) Studies on host resistance of Mungbean against Mungbean Yellow Mosaic Virus in the agro-ecological condition of lateritic zone of West Bengal. Bioscan 8(2):583–587

Poolsawat O, Kativat C, Arsakit K, Tantasawat PA (2017) Identification of quantitative trait loci associated with powdery mildew resistance in Mungbeanusing ISSR and ISSR-RGA Marker. Mol Breed 37(12):150. https://doi.org/10.1007/s11032-017-0753-2

Rana DS, Dass A, Rajanna GA, Kaur R (2016) Biotic and abiotic stress management in pulses. Indian J Agron 61(4th IAC Special issue):S238–S248

Reddy KS (2009) Identification and inheritance of a new gene for powdery mildew resistance in Mungbean (*Vigna radiata* L. Wilczek). Plant Breed 128(5):521–523. https://doi.org/10.1111/j.1439-0523.2008.01609.x

Reddy KS, Pawar SE, Bhatia CR (1994) Inheritance of powdery mildew (*Erysiphe polygoni* DC) resistance in Mungbean (*Vigna radiata* L. Wilczek). Theor Appl Genet 88(8):945–948. https://doi.org/10.1007/BF00220800

Rohilla V, Yadav RK, Poonia A, Sheoran R, Kumari G, Shanmugavadivel PS, Pratap A (2022) Association mapping for yield attributing traits and yellow mosaic disease resistance in Mungbean. Front Plant Sci 12:749439. https://doi.org/10.3389/fpls.2021.749439

Sahoo JP, Samal KC, Lenka D, Beura SK, Behera L, Sawant SB (2022) Cercosporaleaf spot (CLS) disease resistance screening of Mungbean [*Vigna radiata* (L.) Wilczek] germplasms. Biol Forum 14:1115–1119. https://doi.org/10.13140/RG.2.2.19286.11846

Sai CB, Nagarajan P, Raveendran M, Rabindran R, Kannan Bapu JR, Senthil N (2017) Understanding the inheritance of Mungbean yellow mosaic virus (MYMV) resistance in Mungbean (*Vigna radiata* L. Wilczek). Mol Breed 37:63. https://doi.org/10.1007/s11032-017-0650-8

Selvi R, Muthiah AR, Manivannan N, Raveendran TS, Manickam A, Samiyappan R (2006) Tagging of RAPD marker for MYMV resistance in Mungbean (*Vigna radiata* (L.) Wilczek). Asian J Plant Sci (2)5:277–280

Shukla GP, Pandya BP (1985) Resistance to yellow mosaic in greengram. SABRAO J Breed Genet 17:165–171

Singh RA, Gurha SN (2005) Search for sources of resistance to powdery mildew in Mungbean and urdbean. In: Proceedings of the second global conference on plant health-global wealth, MPUAT, Udaipur, India, 25–29 Nov 2005, vol 256. ISMPP

Singh CM, Pratap A, Gupta S, Biradar RS, Singh NP (2020) Association mapping for Mungbean yellow mosaic India virus resistance in Mungbean (*Vigna radiata* L. Wilczek). 3 Biotech 10(2):33. https://doi.org/10.3389/fpls.2021.749439

Subedi S, Neupane S, Ghimire TN (2016) Screening of Mungbean and black gram genotypes as sources of genetic resistance against Mungbean Yellow Mosaic Disease. Nepal J Agric Sci 14:148–155

Sudha M, Karthikeyan A, Nagarajan P, Raveendran M, Senthil N, Pandiyan M, Angappan K, Ramalingam J, Bharathi M, Rabindran R, Veluthambi K (2013) Screening of Mungbean (*Vigna radiata*) germplasm for resistance to Mungbean yellow mosaic virus using agroinoculation. Can J Plant Pathol 35:424–430. https://doi.org/10.1080/07060661.2013.827134

Sudha M, Karthikeyan A, Madhumitha B, Veera Ranjani R, Kanimoli Mathivathana M, Dhasarathan M, Murukarthick J, Samu Shihabdeen MN, Eraivan Arutkani Aiyanathan K, Pandiyan M, Senthil N (2022) Dynamic transcriptome profiling of Mungbean genotypes unveil the genes respond to the infection of Mungbean yellow mosaic virus. Pathogens 11:190. https://doi.org/10.3390/pathogens11020190

Sun F, Sun S, Zhu L, Duan C, Zhu Z (2019) Confirmation of *fusarium oxysporum* as a causal agent of Mungbean wilt in China. Crop Prot 117:77–85. https://doi.org/10.1016/j.cropro.2018.11.017

Sun S, Zhu L, Sun F, Duan C, Zhu Z (2020) Pathotype diversity of *fusarium oxysporum* f. sp. mungcolacausing wilt on Mungbean (*Vigna radiata*). Crop Pasture Sci 71(10):873. https://doi.org/10.1071/CP20164

Tantasawat PA, Poolsawat O, Kativat C, Arsakit K, Papan P, Chueakhunthod W, Pookhamsak P (2022) Inheritance and identification of ISSR-RGA markers associated with powdery mildew resistance in Mungbean for marker-assisted breeding. Chilean J Agric Res 82(1):3–9. https://doi.org/10.4067/S0718-58392022000100003

Taylor JD, Teverson DM, Allen DJ (1996) Identification and origin of races of *pseudomonas syringae*pv. *Phaseolkola* from Africa and other bean growing areas. Plant Pathol 45(3):469–478. https://doi.org/10.1046/j.1365-3059.1996.d01-147.x

Thakur RP, Patel PN, Verma JP (1977) Genetical relationships between reactions to bacterial leaf spot, yellow mosaic and Cercospora leaf spot diseases in Mungbean (*Vigna radiata*). Euphytica 26(3):765–774. https://doi.org/10.1007/BF00021705

Vairam N, Lavanya SA, Muthamilan M, Vanniarajan C (2016) Screening of M3 mutants for yellow vein mosaic virus resistance in greengram [*Vigna radiata* (L.) wilczek]. Int J Plant Sci 11(2):265–269. https://doi.org/10.15740/HAS/IJPS/11.2/265-269

Vani MS, Kumar S, Gulya R (2019) *In vitro* evaluation of fungicides and plant extracts against *fusarium oxysporum*causingwilt of Mungbean. Pharma Innov J 8(8):297–302

Waengwan P, Laosatit K, Lin Y, Yimram T, Yuan X, Chen X, Somta P (2024) A cluster of peronosporaparasitica 13-like (NBS-LRR) genes is associated with powdery mildew (*Erysiphe polygoni*) resistance in Mungbean (*Vigna radiata*). Plants 13(9):1230. https://www.mdpi.com/2223-7747/13/9/1230

Yadav DL, Pandey RN, Jaisani, Gohel NM (2014) Sources of resistance in Mungbean genotypes to cercospora leaf spot disease and its management. Afr J Agric Res 9(41):3111–3114. https://doi.org/10.5897/AJAR2014.8860

Yen JY, Bui TG, Schafleitner R, Lin CY, Huang SM, Chen LF (2016) Marker-assisted selection of disease and pest resistant Mungbeanlinesusing Cel-I genotyping. In: Freyerand B, Tielkes E (eds) Tropentag 2016: solidarity in a competing world-fair use of resources. © CUVILLIER VERLAG, Göttingen, p 82

Yundaeng C, Somta P, Chen J, Yuan X, Chankaew S, Srinives P, Chen X (2020) Candidate gene mapping reveals VrMLO12 (MLO Clade II) is associated with powdery mildew resistance in Mungbean (*Vigna radiata* [L.] Wilczek). Plant Sci. https://doi.org/10.1016/j.plantsci.2020.110594

Yundaeng C, Somta P, Chen J, Yuan X, Chankaew S, Chen X (2021) Fine mapping of QTL conferring Cercospora leaf spot disease resistance in Mungbean revealed TAF5 as candidate gene for the resistance. Theor Appl Genet 134(2):701–714. https://doi.org/10.1007/s00122-020-03724-8

Zhang MC, Wang DM, Zheng Z, Humphry M, Liu CJ (2008) Development of PCR based markers for a major locus conferring powdery mildew resistance in Mungbean (*Vigna radiata*). Plant Breed 127(4):429–432. https://doi.org/10.1111/j.1439-0523.2008.01521.x

Molecular Breeding Approaches for Bruchid Resistance

Changyou Liu

Abstract

The bruchids, *Callosobruchus chinensis* (L.) and *Callosobruchus maculatus* (F.) (Coleoptera: Bruchidae), are among the most destructive storage pests of mungbean (*Vigna radiata* L.). Their infestation leads to substantial quantitative and qualitative losses in stored seeds, severely compromising the grain's nutritional value, viability, and marketability, thereby posing a significant threat to food security and agricultural economics. The application of chemical control strategies is associated with detrimental effects on environmental integrity and human health. Consequently, the development of bruchid-resistant mungbean cultivars through advanced breeding techniques represents not only a more sustainable and eco-friendly strategy but also a major objective in contemporary legume improvement programs. This review highlights the progress in molecular breeding approaches for bruchid resistance in mungbean. Several resistant germplasm resources, including wild species like TC1966 and ACC41, and cultivated varieties such as V2709 and V1128, have been identified. Genetic mapping has revealed key resistance genes, such as the *Br* gene in TC1966 and QTLs like qBr5.1 in ACC41, which are tightly linked to molecular markers, facilitating efficient selection. The characterization of novel alleles in specific genes, particularly *VrPGIP1* and *VrPGIP2* which encode polygalacturonase-inhibiting proteins, has elucidated key aspects of the molecular mechanism conferring bruchid resistance in mungbean. These findings have facilitated the application of marker-assisted selection (MAS), effectively accelerating the development of resistant mungbean cultivars. These advancements not only reduce reliance on chemical controls but also contribute to sustainable agriculture and improved food security. Future efforts should focus on applying these findings in breeding

C. Liu (✉)
Institute of Cereal and Oil Crops, Hebei Academy of Agriculture and Forestry Sciences, Shijiazhuang, Hebei, China

S. Paramasivam et al. (eds.), *Mungbean*,
https://doi.org/10.1007/978-981-95-7637-1_6

programs and exploring new resistance sources to ensure long-term effectiveness.

Keywords

Bruchid resistance · Molecular Breeding · QTL Marker

1 Introduction

Bruchids (Coleoptera: Bruchidae), commonly referred to as bean weevils or seed beetles, are major storage pests that infest grain legumes, including mungbean (*Vigna radiate* L.). Among the species that pose significant threats to stored mungbean are *Callosobruchus chinensis*, *Callosobruchus maculatus*, *Callosobruchus analis*, and *Callosobruchus phaseoli*, which can cause severe quantitative and qualitative losses to the seeds. Among various bruchid species, *Callosobruchus chinensis* and *Callosobruchus maculatus* pose the most significant threat to mungbean during postharvest storage and transport. Infestation by these pests leads to substantial seed mass loss, severely compromises seed viability (germination capacity), and ultimately renders the seeds unsuitable for human consumption, agricultural planting, or commercial use (Talekar and Lin 1992). Although bruchid infestations in stored mungbeans can be controlled through chemical fumigants, such as aluminum phosphide, the adverse effects of these insecticides—including environmental contamination, potential health hazards, and increased economic costs—render the development of host plant resistance a more sustainable and preferable management strategy. Consequently, breeding bruchid-resistant mungbean cultivars has become a major objective in modern legume improvement programs.

2 Germplasm Resources Resistant to Bruchids

Screening and identifying germplasm resources resistant to bruchids is the foundation of mungbean breeding for resistance to bruchids. So far, mainly seven germplasm resources resistant to bruchids have been screened and identified worldwide (Table 1), including three wild species and four cultivated species. TC1966 and ACC41 are two wild mungbean germplasm resources with resistance to bruchids, which have been most frequently utilized in previous studies as resistant materials. In fact, ever since they were screened and identified, attempts have been made to introduce the resistance genes into cultivated mungbean varieties. However, these wild materials often possess undesirable traits in breeding, such as pod shattering, rampant growth, late maturity, and small seed size. Using conventional crossing often requires multiple rounds of hybridization and backcrossing to possibly break the linkage drags between the bruchids resistance genes and these unfavorable traits.

The cultivated mungbean germplasm accessions V1128 and V2817 exhibit complete resistance to both *C. chinensis* and *C. maculatus* (Somta et al. 2008), while the

Table 1 The mungbean germplasm resources resistant to bruchids

Name	Types	Origin	Resistance to *Callosobruchus chinensis*	Resistance to *Callosobruchus maculatus*	References
TC1966	Wild	Madagascar	Complete	Complete	Fujii and Miyazaki (1987), Young et al. (1992), Kaga and Ishimoto (1998)
ACC41	Wild	Australia	Complete	Complete	Lambrides and Imrie (2000), Miyagi et al. (2004)
ACC23	Wild	Australia	Complete	Complete	Lambrides and Imrie (2000)
V2709	Cultivated	India	Moderate	Moderate	Talekar and Lin (1992), Somta et al. (2007)
V2802	Cultivated	Philippines	Moderate	Moderate	Talekar and Lin (1992), Somta et al. (2007)
V1128	Cultivated	India	Complete	Complete	Somta et al. (2008)
V2817	Cultivated	Nigeria	Complete	Complete	Somta et al. (2008)

previously identified sources V2709 and V2802 show a moderate level of resistance (Talekar and Lin 1992), which is attributed to antibiosis factors present in the cotyledons.

2.1 Genetic Mapping of Bruchid Resistance Gene

The availability of molecular markers tightly linked to bruchid resistance genes would facilitate highly efficient, early-generation selection. Consequently, this approach could drastically reduce the reliance on extensive and costly phenotypic screening for resistance, thereby conserving valuable resources within mungbean breeding programs. The resistance to *Callosobruchus chinensis* in the wild mungbean accession TC1966b is conferred by a single dominant gene, *Br*, as demonstrated by Kitamura et al. (1988). Young et al. (1992) employed restriction fragment length polymorphism (RFLP) markers to map the bruchid resistance gene (*Br*) in an F_2 population derived from a cross between the resistant wild mungbean accession TC1966 and the susceptible cultivar VC3890. Genetic mapping assigned the *Br* locus to linkage group VIII, with the closest RFLP marker, pA882, located at a distance of 3.6 cM. Quantitative trait locus (QTL) analysis confirmed that this genomic region represents a major locus, explaining 87.5% of the phenotypic variance in bruchid resistance, thereby demonstrating its pivotal role in conferring the resistant trait. Through fine-scale mapping of the bruchid resistance gene *Br* in the wild mungbean accession TC1966, Kaga and Ishimoto (1998) delineated its position to a 0.7 cM interval flanked by the RFLP markers Bng110 and Bng143. The *Br* locus was found to be tightly linked to the marker Bng143, with a genetic distance of merely 0.2 cM. Miyagi et al. (2004) established the first mungbean bacterial artificial chromosome (BAC) library using the DNA from the resistant wild species ACC41 and the susceptible cultivated variety ATF3640, with the vector pBeloBACII, which has a capacity to cover 3.5 times the size of the mungbean genome. Based on this BAC library, they used RFLP probes and their sub-clones for hybridization to identify two STS markers, STSbr1 and STSbr2, that are linked to the bruchid resistance gene. Using a recombinant inbred line (RIL) population derived from the cross Berken/ACC41, Mei et al. (2009) consistently detected a major QTL conferring bruchid resistance across four distinct environments. This stable QTL, which explained a notable portion of the phenotypic variance (e.g., >10%), was flanked by the molecular markers VrCS155–1 and VrCS161. The bruchid resistance gene in mungbean accession V2802 was initially pinpointed by the expressed sequence tag-simple sequence repeat (EST-SSR) marker DMB-SSR158, which demonstrated perfect co-segregation with the resistant phenotype (Chotechung et al. 2011). Subsequent fine-mapping by Chotechung et al. (2016) delimited the locus to a 38-kb genomic segment on chromosome 5, within which the gene *VrPGIP2*, encoding a polygalacturonase-inhibiting protein (PGIP), was identified as the most probable candidate gene conferring resistance against *Callosobruchus.*

Kaewwongwal et al. (2017) conducted fine-scale mapping of the *Br* locus in the bruchid-resistant mungbean accession V2709, delineating it to a 237.35-kb genomic interval harboring eight annotated genes. Among these, two adjacent genes encoding polygalacturonase-inhibiting proteins, *VrPGIP1* (LOC106760236) and *VrPGIP2* (LOC106760237), were identified and found to possess novel alleles (designated *VrPGIP1-1*and *VrPGIP2-2*, respectively) in V2709. The *VrPGIP2-2* allele is characterized by a single nucleotide polymorphism (G to C) at position 554, resulting in a non-synonymous substitution of proline to arginine at residue 185. In contrast, the *VrPGIP1-1* allele contains 43 SNPs compared to the wild-type sequence, 20 of which cause amino acid alterations, including a threonine to proline change at the homologous residue 185. Functional validation through heterologous expression and purification confirmed that both VrPGIP1 and VrPGIP2 proteins exhibit significant polygalacturonase-inhibiting (PG-inhibiting) activity. In an artificial seed assay, the incorporation of either protein conferred a strong, dose-dependent resistance against bruchid infestation, likely through the inhibition of insect-derived polygalacturonases (PGs) critical for larval digestion and development (Zhang et al. 2021).

The bruchid resistance locus (*Br*) in the cultivated mungbean accession V1128 was mapped to an approximately 288 kb genomic interval on chromosome 5, flanked by the markers DMB158 and VRBR-SSR033 (Liu et al. 2018). In a parallel investigation of the wild mungbean accession ACC41, Kaewwongwal et al. (2020) delineated the genetic basis of resistance, identifying a major quantitative trait locus (qBr5.1) and a minor QTL (qBr5.2) separated by only 0.3 cM. The major locus, qBr5.1, which accounted for approximately 82% of the resistance phenotypic variation in an F_2 population, was fine-mapped to a 237.35 kb region on chromosome 5. This region harbors eight annotated genes, including the candidate genes *VrPGIP1* and *VrPGIP2*. Subsequent sequence analysis of ACC41 revealed a novel allele of *VrPGIP1*, designated *VrPGIP1-2*. This allele carries five single nucleotide polymorphisms (SNPs) compared to the wild-type sequence, four of which result in amino acid substitutions at residues 125, 129, 188, and 336. Protein structure analysis indicated that the substitutions at residues 125 and 129, located within the β-sheet B1 region—a domain critical for interaction with insect polygalacturonase (PG)—are likely central to the enhanced bruchid resistance by potentially strengthening the inhibition of PG activity. These findings strongly implicate the *VrPGIP1-2* allele as a key determinant of bruchid resistance in the wild accession ACC41.

3 Conclusions

The development of bruchid-resistant mungbean cultivars is crucial for reducing post-harvest losses and enhancing the economic viability of mungbean production. Molecular breeding approaches have significantly advanced our understanding of the genetic basis of bruchid resistance in mungbean. The identification and mapping of major resistance loci, such as the *Br* gene in the wild accession TC1966 and the qBr5.1 QTL in ACC41, have enabled more efficient, targeted precision breeding strategies for bruchid resistance in mungbean. The subsequent development of closely linked molecular markers has significantly streamlined the selection process, reducing reliance on extensive and costly phenotypic screening and thereby lowering the overall time and financial investment in cultivar development.

Moreover, the characterization of new alleles, such as *VrPGIP1-1*, *VrPGIP2-2*, and *VrPGIP1-2*, has provided deeper insights into the molecular mechanisms underlying bruchid resistance. The identification of these genetic loci provides a robust foundation for breeding bruchid-resistant mungbean cultivars. Integrating these molecular tools with conventional breeding programs constitutes a strategic pathway for rapidly developing resilient varieties. This approach not only mitigates postharvest losses caused by bruchid infestations, thereby enhancing food security, but also significantly reduces the dependence on chemical pesticides, promoting more sustainable agricultural practices.

In conclusion, the collective progress in elucidating the genetic basis of bruchid resistance and applying precision breeding strategies, such as marker-assisted selection, represents a pivotal advancement toward achieving sustainable mungbean production. This integrated approach not only facilitates the development of durable, resistant cultivars but also significantly reduces reliance on chemical pesticides, thereby enhancing global food security and agricultural sustainability. Future efforts should focus on the practical application of these findings in breeding programs and the continued exploration of novel resistance sources to ensure the long-term effectiveness of resistance strategies.

References

Chotechung S, Chankaew S, Srinives P, Somta P (2011) Identification of DNA markers associated with bruchid resistance in mungbean. Khon Khan Agric J 39:221–226

Chotechung S, Somta P, Chen J, Yimram T, Chen X, Srinives P (2016) A gene encoding a polygalacturonase-inhibiting protein (PGIP) is a candidate gene for bruchid (Coleoptera: bruchidae) resistance in mungbean (*Vigna radiata*). Theor Appl Genet 129:1673–1683

Fujii K, Miyazaki S (1987) Infestation resistance of wild legumes (*Vigna sublobata*) to adzuki bean weevil, *Callosobruchuschinensis* (L.) (Coleoptera: Bruchidae) and its relationship with cytogenetic classification. App Entomol Zool 22:229–230

Kaewwongwal A, Chen J, Somta P, Kongjaimun A, Yimram T, Chen X, Srinives P (2017) Novel Alleles of two tightly linked genes encoding polygalacturonase-inhibiting proteins (VrPGIP1 and VrPGIP2) associated with the Br Locus that confer bruchid (*Callosobruchus*spp.) Resistance to Mungbean (*Vigna radiata*) accession V2709. Front Plant Sci 8:1692

Kaewwongwal A, Liu CY, Somta P, Chen JB, Tian J, Yuan XX, Chen X (2020) A second VrPGIP1 allele is associated with bruchid resistance (*Callosobruchus spp.*) in wild mungbean (*Vigna radiata* var. *sublobata*) accession ACC41. Mol Genet Genomics 295:275–286

Kaga A, Ishimoto M (1998) Genetic localization of a bruchid resistance gene and its relationship to insecticidal cyclopeptide alkaloids, the vignatic acids, in mungbean (*Vigna radiata* L. Wilczek). Mol Gen Genomics 258:378–384

Kitamura K, Ishimoto M, Sawa M (1988) Inheritance of resistance to infestation with adzuki bean weevil in *Vigna sublobata* and successful incorporation to *V. radiata*. Japan J Breed 38:459–464

Lambrides CJ, Imrie BC (2000) Susceptibility of mungbean varieties to the bruchid species *Callosobruchus maculatus* (F.), *C. phaseoli* (Gyll.), *C. chinensis* (L.), and *Acanthoscelidesobtectus* (Say.) (Coleoptera: Chrysomelidae). Aust J Agric Res 51:85–90

Liu CY, Su QZ, Fan BJ, Cao ZM, Zhang ZX, Wu J, Cheng XZ, Tian J (2018) Genetic mapping of bruchid resistance gene in V1128. Acta Agron Sin 44(12):1875–1881

Mei L, Cheng XZ, Wang SH, Wang LX, Liu CY, Sun L, Xu N, Humphry ME, Lambrides CJ, Li HB, Liu CJ (2009) Relationship between bruchid resistance and seed mass in mungbean based on QTL analysis. Genome 52:589–596

Miyagi M, Humphry M, Ma Z, Lambrides CJ, Bateson M, Liu CJ (2004) Construction of bacterial artificial chromosome libraries and their application in developing PCR-based markers closely linked to a major locus conditioning bruchid resistance in mungbean (*Vigna radiata* L. Wilczek). Theor Appl Genet 110:151–156

Somta P, Ammaranan C, Ooi P, Srinives P (2007) Inheritance of seed resistance to bruchids in cultivated mungbean (*Vigna radiata*, L. Wilczek). Euphytica 155:47–55

Somta C, Somta P, Tomooka N, Ooi AC, Vaughan DA, Srinives P (2008) Characterization of new sources of mungbean (*Vigna radiata* (L.) Wilczek) resistance to bruchids, *Callosobruchus* spp. (Coleoptera: Bruchidae). J Stored Prod Res 44:316–321

Talekar NS, Lin CL (1992) Characterization of *Callosobruchus chinensis*(Coleoptera: Bruchidae) resistance in mungbean. J Econ Entomol 85:1150–1153

Young ND, Kumar L, Menancio-Hautea D, Danesh D (1992) RFLP mapping of a major bruchid resistance gene in mungbean (*Vigna radiata*, L. Wilczek). Theor Appl Genet 84:839–844

Zhang QX, Yan Q, Yuan XX, Lin Y, Chen JB, Wu RR, Xue CC, Zhu YL, Chen X (2021) Two polygalacturonase-inhibiting proteins (VrPGIP) of *Vigna radiata* confer resistance to bruchids (Callosobruchus spp.). J Plant Physiol 258–259:153376

Molecular Breeding Approaches for Abiotic Stress Tolerance in Mungbean (*Vigna radiata* L. Wilczek)

Sivakumar Paramasivam, Kuttralingam Sasikala, Jothi Kanmani Bharathi, Subramanian Rajesh, and Muthiyan Pandiyan

Abstract

Mungbean (*Vigna radiata* L. Wilczek) is a significant short-duration legume crop prized for its excellent nutritional content, has capacity to fix atmospheric nitrogen, and plays a major role in sustainable agriculture. Abiotic stresses like heat, salt, drought, and nutrient shortages, which are predicted to worsen due to climate change, significantly limit its productivity. Enhancing mungbean's ability to withstand abiotic stress is a primary breeding objective; yet, the crop's low genomic resources, restricted genetic base, and its intricate, polygenic nature of stress tolerance traits impede success. Recent developments in genomics, proteomics, genome editing, metabolomics, genome-wide association studies (GWAS), quantitative trait loci (QTL) mapping, and molecular breeding have improved our knowledge on the mechanisms underlying the stress tolerance and made it easier to identify important candidate genes and biomarkers.

Keywords

Vigna radiata · Abiotic stress tolerance · Molecular breeding · QTL mapping · GWAS · Transcriptomics · Genome editing · Multi-omics

S. Paramasivam (✉) · K. Sasikala · M. Pandiyan
Dr. M.S. Swaminathan Agricultural College and Research Institute, Tamil Nadu Agricultural University, Eachangkottai, Thanjavur, Tamil Nadu, India
e-mail: sivakumar.p@tnau.ac.in

J. Kanmani Bharathi
Faculty of Agriculture, Annamalai University, Annamalainagar, Tamil Nadu, India

S. Rajesh
Department of Plant Biotechnology, CPMBB, Tamil Nadu Agricultural University, Coimbatore, Tamil Nadu, India

S. Paramasivam et al. (eds.), *Mungbean*,
https://doi.org/10.1007/978-981-95-7637-1_7

1 Introduction

Mungbean (*Vigna radiata*) is a short-duration leguminous plant cultivated in South and Southeast Asia (Diatta et al. 2024). However, the plant's productivity is limited by abiotic stresses like drought, salinity, and heat, especially under rainfed and marginal land conditions. The increased frequency and intensity of these stresses due to climate change threaten the sustainable production of mungbean. Traditional breeding methods to develop stress-tolerant plants had limited success due to the narrow genetic diversity of cultivated mungbean because stress tolerance is a complex trait regulated by many genes (Manu et al. 2022). To overcome these limitations, molecular approaches can significantly assist mungbean improvement by elucidating the genetic basis of stress tolerance and facilitating precise trait modification. Understanding the biochemical responses of mungbean to various abiotic stresses is gaining attention on for development of stress tolerant varieties.

Molecular breeding employs advanced genomic tools such as marker-assisted selection (MAS), genomic selection, and genome-wide association studies (GWAS) to identifies and incorporates desirable alleles associated stress tolerance. These approaches enable precise and efficient selection of stress-resilient genotypes by linking phenotypic traits with specific genetic markers, thereby accelerating the breeding process compared to conventional methods. This chapter aims to analyze the recent progress in molecular approaches targeting mungbean for enhanced abiotic stress tolerance.

2 Overview of Abiotic Stresses in Mungbean

Mungbean crop is susceptible to several abiotic stresses that can have significant effect on plant physiology, metabolism, and ultimately yield. These include drought, salt, heat, cold, and water logging (submergence). Mungbean crop is generally cultivated in tropical, subtropical, and temperate regions of the world.

Mungbean growth and productivity are highly sensitive to unfavorable environmental conditions. Drought and high temperatures are part of abiotic stressors that have severe physiological, morphological, and reproductive effects which are reflected in high losses of productivity and seed quality (Hanif and Wahid 2018). Reproductive development is the most susceptible stage, with even minor changes in the availability of water having an ability to reduce the seed production. Salinity further exacerbates these effects by impairing crop performance through its influence on both plants and their mutualistic symbiont, especially *Rhizobium* spp., which play a critical role in biological nitrogen fixation (BNF) (Kumar et al. 2020). Even relatively low salinity levels, i.e., 50 mM sodium chloride, have typically caused mungbean yields to be reduced by more than 60 percent (Desai et al. 2023). In addition, germination, vegetative growth, flowering, and pod filling stages are very susceptible and this reduces overall yield (Sehrawat et al. 2015). Heat stress occurring during pod filling stage severely affects the pod and grain filling development (HanumanthaRao et al. 2016; Kaur et al. 2015). Baroowa et al. (2016) found that the water stress in mungbean can reduce photosynthesis rate due to degeneration of chlorophyll and closure of stomata.

During drought condition, the crop growth, flower formation, and pod development of mungbean were gradually affected (Chaudhary et al. 2022). The elevated temperatures (above optimal ranges) abort flowers and pods, reduce seed size, and decrease the overall yield. Salinity stress occurs when high concentrations of soluble salts in the soil adversely affect mungbean growth and development (HanumanthaRao et al. 2016). The crop is sensitive to temporary water logging or submergence conditions in the soil at any stage of growth (Nair et al. 2012). Prolonged exposure to excess water has been shown to impair root function, lower chlorophyll content in leaves, and significantly reduced seed yield (Islam et al. 2008). Mungbean plants respond to abiotic stresses through various mechanisms such as the accumulation of osmolytes (e.g., proline), antioxidant enzyme activities, and morphological adaptations like root system modification. Breeding for new submergence tolerant genotypes is a major breeding goal to enhance the adaptability of mungbean genotypes in various climatic conditions (Islam et al. 2007; Shukla et al. 2019). Multiple concurrent situations, such as heat and drought, are significant obstacles to mungbean productivity. With climate change, this is anticipated to become more often. To the best of our knowledge, mungbean has not been the subject of studies assessing several concurrent abiotic stressors throughout its reproductive phase. Determining the essential physiological processes involved in acclimatization to various abiotic stress situations and any possible alternatives involved requires immediate molecular breeding investigation.

2.1 Drought Stress Tolerance in Mungbean

Drought tolerance in plants is a multifaceted quantitative trait that arises from the interplay of numerous metabolic pathways and the regulation of stress-responsive genes (Bangar et al. 2019).

2.1.1 Factors Influencing Improved Drought Stress Tolerance in Mungbean

Drought stress has detrimental impacts on the morpho-physiological and biochemical processes and structures of plant development as well as molecular regulation and reduces grain yield in mungbean. Experimental evidence amends that drought at both the vegetative and reproductive phases significantly reduces photosynthetic competence and gaseous-exchange performance (Moradi et al. 2008; Baroowa et al. 2016; Jisha and Puthur 2016; Bangar et al. 2019; Nivethitha et al. 2024). The predominant effect of mild water deficit on photosynthetic performance is in the form of stomatal limitations, whereas at higher soil water deficit, photosynthetic efficiency and leaf gas exchange are also limited by non-stomatal factors. Furthermore, gas exchange characteristics as measured during stress condition and after stress are still beneficial measures of genotypic adaptation to drought. As an example, Dutta et al. (2016) have reported that the mungbean genotype K-851 had better tolerance at the seedling stage, which is mainly due to its effective leaf water balance and its balanced chlorophyll levels. Photo-oxidative damages, transpiration, photochemical efficiencies are hence commonly used as stable screening parameters to exhibit drought tolerance by plant breeders and physiologists (Narina et al. 2014). Ahmad et al. (2015) demonstrated that drought stress significantly reduces the yield of mungbean, with the extent of reduction varying according to irrigation frequency (3, 5, 7, and 9 days) and water availability. Seed yield declined across all treatments, but the most severe reduction occurred under a 9-day irrigation interval, dropping from 1146 to 78 kg/ha. Similarly, Jisha and Puthur (2016) reported that seed priming with β-amino butyric acid (BABA at 0.5–2.5 mM) in mungbean genotypes such as Pusa Ratna, Pusa 9531, and Pusa Vishal influenced plant responses to drought stress. Priming of seeds has been reported to promote physiological and biochemical characteristics that promote drought tolerance in mungbean. Primed seeds had elevated chlorophyll fluorescence, mitochondrial activity and photosynthetic performance, and seedling growth properties, e.g., shoot length and biomass. They also demonstrated lower malondialdehyde (MDA) accumulation, higher proline and total soluble sugar contents, enhanced nitrate reductase activity, and increased antioxidant enzyme activities such as superoxide dismutase (SOD) and guaiacol peroxidase. In parallel, using polyamines (mixtures of putrescine, spermidine, and spermine) as primers reduced membrane damage and augmented proline, soluble proteins, and sugars as a way of promoting better growth and yield during drought stress (Sadeghipour 2019). Alderfasi et al. (2017) also tested two mungbean genotypes, Kawmay-1 and VC2010, emphasizing that stomatal conductance, shoot water content, chlorophyll concentration, and antioxidant photoprotective enzymes are some of the key traits that relate to drought-tolerance.

Bangar et al. (2019) examined 25 mungbean genotypes under drought conditions during vegetative and reproductive stages and reported significant reductions in leaf area, membrane stability index (MSI), proline levels in leaves, relative water content (RWC), and height and plant elevation. Researchers have reported that isozyme systems, such as ascorbate (AsA), ascorbate glutathione (AsA GSH), cytosolic ascorbate peroxidase (cAPX), dehydroascorbate reductase (DHAR), and superoxide dismutase (SOD), are the key factors in drought tolerance and serve as useful biochemical indicators to distinguish between genotypic variations in mungbean (Sanghani et al. 2015). Moreover, Alderfasi et al. (2017) also reported a dramatic upregulation of antioxidant enzymes such as catalase, polyphenol oxidase, superoxide dismutase, and ascorbate peroxidase during water stress. The increased expression alleviates oxidative injury due to the fact that it downregulates reactive oxygen species (ROS), thus contributing to drought tolerance. Ali et al. (2018) examined seed germination characters and tested seed germination ability under drought stress in mungbean germplasm and found an increase in soluble protein and malondialdehyde (MDA) content during the water scarce environment. Similarly, activities of antioxidant enzymes, peroxidase, superoxide dismutase, catalase, ascorbate peroxidase, and total flavonoid content levels escalated significantly in response to drought, as compared to a reduction in ascorbic acid (AsA) and total phenolic content (TPC) when stress was high. These biochemical and physiological changes can serve as trustworthy criteria to come up with drought-resistant varieties of mungbean. In a previous study, Kumari et al. (2016) arrived at a similar decision and indicated that drought-stressed mungbean did have a higher accumulation of proline and a lower osmotic injury.

The mungbean genotype AU-M4 is an overproducer variant of plant growth-promoting rhizobacteria (PGPR) and have shown to have high drought tolerance. Typically, AU-M4 strain displays inducible ACC deaminase action, great amounts of indole-3-acetic acid, and excellent levels of inorganic phosphate solvents when compared to the wild-type strains. This type has an increased biomass of plants, high proline levels, water retention capacity, and decreased injury of the osmotic stress under drought stress. Additionally, drought tolerance toward mungbean was also enhanced upon inoculation of osmoprotectant rhizobacterial isolates, A124-K and Ver5-K, which had the possibility of glycine betaine production (Maryani et al. 2018). The cultivars of the mungbean with a drought tolerance, e.g., BARI Mung-8, BMX-08010-2, BMX-010015, and BMX-08009-7, have shown increased proline accumulation and maintenance of chlorophyll pigmentation during flowering and pod maturation under water-stressed conditions (Islam et al. 2022). Meanwhile, such genotypes have exhibited better water-relationship parameters, such as high relative water content (RWC) and water retention capacity (WRC), lower water saturation deficit (WSD) and water uptake capacity (WUC). Nivethitha et al. (2024) also found genotypes VBN3, VBN4, COGG22-01, and VGG17-004, which maintained high RWC (72.2176.83%) and reduced transpiration as a result of stomata and leaf structural adjustments. These genotypes also stored high proline (0.54–0.67 mg/g) and soluble sugars (19.56–24.12 mg/g dry weight), thus facilitating the occurrence of effective osmotic adjustment to drought stress.

2.1.2 Genomic Approaches for Drought Tolerance in Mungbean

Genomic approaches are very limited for identification of drought-resistant candidate genes, transcription factors, and/or QTLS in mungbean (Liu et al. 2017; Labbo et al. 2018; Fatima et al. 2020). The QTLS namely qPH5A and qMLA2A were consistently identified across four different environments. (Liu et al. 2017).

Drought stress has been associated with the downregulation of proline dehydrogenase, which contains a 118 bp intron. Interestingly, the higher seed yield of the low leaf water loss (LWL) genotype, despite exhibiting a hotter canopy, was linked to maximum net assimilation and yield under drought conditions (Raina et al. 2016). At the molecular level, Labbo et al. (2018) identified 71 AP2/ERF transcription factors in mungbean, classified into AP2 (16), ERF (22), RAV (2), and DREB (30) families. Among these, the DREB subgroup plays a particularly crucial role in drought adaptation. Notably, five DREB genes (VrDREB5, VrDREB12, VrDREB13, VrDREB22, and VrDREB30) exhibited strong upregulation under drought stress, making them promising candidates for the genetic improvement of drought tolerance in mungbean.

Eight SnRK2 genes were identified for osmotic stress tolerance through genome-wide analyses. Among the eight genes, VrSnRK2.6c exhibit higher expressions under drought stress condition. The incorporation and expression of all these

VrSnRK2 genes in the mungbean genotypes lead to the development of drought tolerance (Fatima et al. 2020). Twelve SnRK2 genes were identified in a related species *Vigna unguiculata,* and in silico analysis characterized their chromosomal distribution and evolutionary relationships, highlighting their similarity to stress-responsive homologs in other crop species (Rajesh et al. 2025). The genome-wide analysis of WRKY 10 and other transcription factors in mungbean identified 213 transcription factors, including two WRKY transcription factors (Vradi05g21980 and Vradi0158s00480) that are involved in developing tolerance to heat, drought, and osmatic stress (Zandalinas et al. 2018). Chang et al. (2023) resequenced 282 mungbean accessions to enumerate a high-resolution variant map and later developed genome-wide association studies (GWAS) to outline the location of the loci related to drought tolerance. The findings have recognized 146 single-nucleotide polymorphisms (SNPs) highly associated with drought-responsive traits among which 26 multi-trait candidate loci were scrutinized. In these regions, 215 candidate genes were described of which 11 were transcription regulators, 7 protein kinase genes, and a number of other protein coding genes that could be involved in pathways of drought stress response.

2.2 Developing Salinity Stress Tolerance in Mungbean

Mungbean are generally susceptible to salt stress. Salinity stress seriously caused the yield losses in mungbean particularly during flowering and pod development stages. Mungbean genotypes exhibited various degrees of tolerance and/or resistance against salinity stress. There are few reports available for the development of salinity tolerance varieties in mungbean (HanumanthaRao et al. 2016; Sehrawat et al. 2019; Kobir et al. 2022; Ahmed et al. 2024; Deeroum et al. 2024).

2.2.1 Factors Influencing Improved Salinity Stress Tolerance in Mungbean

Mungbean (*Vigna radiata* L.) production is greatly limited by soil salinity, especially in the coastal areas where it is one of the cultivated rice fallow crops. The high costs of yield losses due to salinity stress have spurred small-scale efforts to incorporate abiotic stress tolerance to mungbean breeding programs. Complexity of salinity-related stress and lack of sufficiently powerful and effective screening methods so far have hindered significant development of recognitions of tolerant genotypes (Singh and Singh 2011; Kumar et al. 2012).

Salt stress severely affects the mungbean crop by reducing the seed germination, yield, and yield-related biometrical traits (Promila and Kumar 2000; Rabie 2005; Ahmed 2009). BARI Mung4 variety has demonstrated better survival at higher sodium chloride concentration, particularly a yield-contributing character (Naher and Alam 2010). Kandil et al. (2012) found that the increasing salinity concentrations significantly reduced various germination factors including shoot and root length in Kawmy-1 and IV 2010 mungbean varieties. Kumar et al. (2012) found that the hybrid from VMGG1 and VBN1 cross exhibited reduced transpiration rate and stomatal conductance (2012). A very meager progress has been made so far for developing salinity tolerant genotypes in mungbean (Ambede et al. 2012; HanumanthaRao et al. 2016). Rahman et al. (2016) found that the BINA mung 8 performed better in terms of shoot and root growth at high 50 mM NaCl salt concentration. Manasa et al. (2017) identified few salt-tolerant genotypes namely ML1299, EC 693357, EC 693358, EC 693366, and EC 693371 at both seedling and plant development stages. Hapsari and Trustinah (2018) reported that Murai and Vima 1 mungbean genotypes are tolerant to salt stress and exhibited the significant level of seed germination and root deployment.

Under salinity stress, plants exhibit significant alterations in protein expression, including the upregulation of photosynthetic proteins, enzymes involved in reactive oxygen species (ROS) scavenging, LEA proteins, membrane-associated proteins, osmolyte biosynthesis, and those linked to carbohydrate metabolism (Arif et al. 2020). Mankar et al. (2021) further reported that proteins, secondary metabolites, osmolytes, and antioxidants increased under mild salt stress but declined at higher salinity levels. In contrast, photosynthetic pigments decreased and membrane damage intensified as salt concentrations increased. Notably, mungbean varieties such as 'PKU-AKM 12–28' and 'PKU-AKM 8802' exhibited relatively better tolerance, being the least affected by NaCl stress.

Kobir et al. (2022) found that the mungbean genotypes BMXKI-112009-21, MMATV07, BARI Mung-6, and BMMP-201524 exhibited tolerance to salt stress at electrical conductivity of 8 dSm-1. Similarly, Ahmed et al. (2024) found that the mungbean varieties like Abbas Mung, NM-2021, NM-92, NM-19-19, and NM-98 are tolerant to salinity stress. They also observed no reduction in germination, and shoot and root length at different salt concentration levels. Deeroum et al. (2024) crossed the salinity-tolerant wild mungbean (*Vigna radiata* var. sublobata) with mungbean. Their findings suggested that salt tolerance in mungbeen is governed either by a single recessive gene or by two recessive genes exhibiting an inhibitory interaction.

2.2.2 Genomic Approaches for Salinity Tolerance in Mungbean

Sehrawat et al. (2014) identified 124 putative alleles using SSR markers, which exhibited polymorphic between wild and cultivated species. Very limited salt tolerant alleles were identified and/ or detected in mungbean through GWAS.

Breria et al. (2020) identified five SNPs related to salt stress (50 mM NaCl) tolerance for germination rate. The salt tolerance genes namely Vradi09g09510, Vradi09g09600, and Vradi07g01630 are present in two QTLs. Vradi07g01630 located on chromosome 7 encodes an ammonium transport protein, which plays significant contribution to plants response to salt stress. On the other hand, the Vradi09g09510 and Vradi09g09600 proteins located on chromosome 9 are associated with slat stress tolerance in mungbean

Liu et al. (2022) identified 3 and 4 QTLS related to slat concentration at 100 mM NaCl. They also found that the *VrFRO8* gene reduces the SOD content by adjusting the Fe^{2+}/Fe^{3+} ratio under salt stress. Xu et al. (2023) identified alkali-tolerant genetic loci and candidate genes in mungbean accessions during germination. They discovered 19 alkali-tolerant QTLs comprising SNPs across nine chromosome. These QTLs showed a phenotypic variance between 3.6% and 14.6%, providing valuable perception into the genetic basis of alkali tolerance in mungbean and potential targets for breeding programs.

2.3 Developing Heat Stress Tolerance in Mungbean

Breeding for high-yielding varieties with heat stress tolerance is the primary objective of mungbean crop improvement. The mechanisms responsible for heat stress tolerance in mungbean are complex and involve various physiological and biochemical processes (Bhardwaj et al. 2023). The study of these heat stress mechanisms in mungbean is indeed essential for developing effective strategies to enhance heat tolerance and improve crop resilience.

2.3.1 Factors Influencing Improved Heat Stress Tolerance in Mungbean

Research on mungbean demonstrates repeatedly that the heat stress significantly affects the reproductive stage of this food crop, resulting in significant losses of yield (Kumar et al. 2013; Kaur and Nayyar 2015; Sharma et al. 2016; Priya et al. 2020). It has been reported that the pollen loses viability, fertilization is poor, excess flowers are aborted, and grain-filling is cut short in extreme heat (HanumanthaRao et al. 2016). Mungbean grows and develops best within the temperature range of 283 °C and seed development occurs past 3335C. An increase of 1 °C higher than optimum temperature only would mean a decline of 3540% seed yield in comparison to the plants grown under optimum conditions (Sharma et al. 2016). Noticeably, some of the mungbean germplasms are able to endure high ascending temperatures, growing its seeds as low as 20 °C and 45 °C. Such diversity in heat tolerance is a very precious feature to breed those high-yielding resilient cultivars. The trait of heat tolerance is complex, though, and different genotypes exhibit varying physiological and molecular processes that allow them to adapt (Bhardwaj et al. 2023). Heat stress has negative impacts on plant physiology by reducing the concentration of chlorophyll, the efficiency of photosynthesis, and transpiration and by suppressing the canopy temperature under canopy modulating static control. High temperatures in mungbean are linked with decreasing leaf area and modified stomatal dynamics, which results in a significant decrease in carbon dioxide assimilation and photosynthesis in the vegetative stage (HanumanthaRao et al. 2016). This adverse effect is exaggerated by increased atmospheric CO_2 that leads to the closing of stomata and inhibits the exchange of gasses, thereby reducing the efficiency of photosynthesis. These two factors work synergistically to suppress mungbean growth and development primarily under conditions of elevated temperature and high concentration of CO_2 (Reardon and Qaderi 2017). Further research data show that carbon dioxide incorporation declines considerably at 40 °C which directly impairs photosynthetic activity (Karim et al. 2003). Chand et al. (2018) have found that high-temperature could reduce the concentration of chlorophyll and carotenoids and decrease the chlorophyll stability index in mungbean. The greatest negative impact on leaf conductance was exhibited in the pre-flowering stage of leaves, which were exposed to moderately high temperatures (36-degrees Celsius), with a slight variation in transpiration rates through stages (Islam 2015). The photosynthesis also reduced steadily due to heat stress. Chand et al. (2018) documented the decrease in chlorophyll and carotenoids, together with a decline in the chlorophyll stability index of the prominent mungbean varieties (e.g., MH 421, MH 318, and Basanti) under high temperature conditions. High temperatures affect the death of mungbean biochemical reactions significantly. Rise in ambient temperatures intensifies the production of reactive oxygen species (ROS), namely, hydroxyl radicals, singlet oxygen, superoxide radicals, and hydrogen peroxide. These ROS help prolong the process of degenerating proteins, defecting the membranes, incompetence in the various enzymes, and the eventual oxidative stress. Mansoor and Naqvi (2013) have conducted a study on four genotypes of mungbean namely NCM 89, NM 20-21, NM 121-123, and NM 19-19 where it was indicated that a substantial amount of lipid peroxidation destruction was recorded in the seedling stage of plants which is indicated by the elevated levels of malondialdehyde (MDA). Heat stress also distorts assimilate partitioning and lowers phloem movement between apoplasts and symplasts, thereby reducing accumulation of carbohydrates and pollen viability (Kaur et al. 2015; Hanif and Wahid 2018). Mungbean response to high temperatures is production of heat shock proteins (HSPs). The heat shock proteins play the role of stabilizing proteins. Li et al. (2019) characterize Hsf genes of mungbean and prove the dynamic regulation of Hsf genes during stressful conditions, which provide evidence of revising the Hsf genes of VrHsf to increase the tolerance to abiotic stresses.

2.3.2 Genomic Approaches for Heat Stress Tolerance in Mungbean

Genomic approaches are very limited for heat stress tolerance in mungbean. During heat stress condition, some LEA genes namely VrLEA-55, VrLEA-47, VrLEA-40, and VrLEA-2 were identified. The regulation of these LEA genes under heat stress was reported by Singh et al. (2022).

2.4 Developing Submergence Tolerance in Mungbean

Water logging (submergence) tolerance is a highly complex trait, involving multiple morphological and physiological adaptations that plants have evolved to cope up such stress. One key tolerance mechanism is the production of adventitious roots, which enhance oxygen uptake and improve plant survival under waterlogged conditions. Water logging represents a severe abiotic stress in mungbean (*Vigna radiata* L.) cultivation which adversely affecting growth, development, and yield. Understanding and enhancing these tolerance mechanisms are crucial for improving mungbean resilience in flood-prone environments.

2.4.1 Factors Influencing Improved Submergence Tolerance in Mungbean

Waterlogging during plant growth and development stages drastically decreased dry matter accumulation in the mungbean cultivars KPS1 and CNXP-49, both during the stress period and afterward. This reduction indicates that waterlogging adversely affects overall growth and biomass production in this cultivars, potentially reducing their yield and productivity (Ahmed et al. 2024). Mungbean crops do not survive in the water logging or submergence condition, especially during the early and vegetative stages of growth. They also found that reduced oxygen concentration occurs in the roots grown in the submergence condition (Singh and Singh 2011). The mungbean genotypes T 44 and MH–96–1 have demonstrated tolerance to water logging conditions. These genotypes maintain higher levels of photosynthetic activity, chlorophyll content, and carotenoids, which support their growth and physiological stability under such stress. Additionally, they exhibit enhanced membrane stability and rapid recovery of photosynthesis after water logging stress (Kumar et al. 2012). Waterlogging primarily inhibits photosynthesis, which consequently results in substantial reductions in grain yield (Ahmed et al. 2024; Noble et al. 2020). Water logging during the pod-filling stage led to a substantial reduction in stomatal conductance and photosynthetic rate, decreasing them to approximately 27% and 25% of the control levels, respectively. Despite this stress, the mungbean genotype AVTMB#3 produced significantly larger seeds compared to other genotypes, including the prominent Australian variety Jade-AU. This suggests that AVTMB#3 possesses some resilience to water logging stress, maintaining seed size even under adverse conditions (Ikram et al. 2022). Kyu et al. (2024) reported that mungbean germplasm showed considerable variation in tolerance to transient waterlogging stress across contrasting geographic origins, with significant differences in most traits except root dry mass. Within their mini-core collection, they identified several promising genotypes exhibiting adaptive traits: lines with high emergence (AGG325612, AGG325667, AGG325523), enhanced root dry mass production (AGG325758, AGG325732, AGG325695, AGG325734, AGG325761), and prolific development of adventitious roots (AGG325560, AGG325593, AGG325733, AGG325634, AGG325718, AGG325514, AGG325726). These genotypes possess critical traits linked to waterlogging adaptation and therefore represent valuable parental sources for breeding programs aimed at improving tolerance in mungbean. GWAS analysis revealed that the total dry mass of mungbean plants under waterlogged condition is associated with the genes encoding MORC family CW-type zinc finger protein 3 and zinc finger protein 2B (Kyu et al. 2024).

3 Conclusion

Production of mungbean (*Vigna radiata*) is severely hampered by abiotic stress, especially in light of climate change and growing environmental unpredictability. However, novel approaches to improving stress tolerance in this significant legume crop have been made possible by recent developments in molecular technology.

Methods including genome editing, transcriptomics, proteomics, metabolomics, QTL mapping, GWAS, and molecular marker-assisted selection have all contributed significantly to our understanding of the molecular mechanisms and genetic structure underlying stress tolerance. Combining multi-omics methods, artificial intelligence, and high-throughput phenotyping platforms improves our ability to accurately and efficiently identify genotypes that are stress-resilient. However, issues including low genetic diversity, weak phenotyping systems, and gaps in multi-environment validation continue to be major problems. Utilizing untapped wild relatives and landraces for novel alleles, as well as integrating traditional breeding with state-of-the-art molecular technologies, should be the top priority of future mungbean development initiatives.

References

Ahmad A, Selim MM, Alderfasi AA, Afzal M (2015) Effect of drought stress on mung bean (*Vigna radiata* L.) under arid climatic conditions of Saudi Arabia. Ecosyst Sustain Dev 192:185–193

Ahmed S (2009) Effect of soil salinity on the yield and yield components of mungbean. Pak J Bot 41(1):263–268

Ahmed SR, Asghar MJ, Hameed A, Ghaffar M, Shahid M (2024) Advancing crop improvement through GWAS and beyond in mung bean. Front Plant Sci 15:1436532

Alderfasi AA, Alzarqaa AA, AL-Yahya FA, Roushdy SS, Dawabah AA, Alhammad BA (2017) Effect of combined biotic and abiotic stress on some physiological aspects and antioxidant enzymatic activity in mungbean (*Vigna radiata* L.). Afr J Agric Res 12(9):700–705. https://doi.org/10.5897/AJAR2016.12084

Ali Q, Javed MT, Noman A, Haider MZ, Waseem M, Iqbal N, Waseem M, Shah MS, Shahzad F, Perveen R (2018) Assessment of drought tolerance in mung bean cultivars/lines as depicted by the activities of germination enzymes, seedling's antioxidative potential and nutrient acquisition. Arch Agron Soil Sci 64(1):84–102

Ambede JG, Netondo GW, Mwai GN, Musyimi DM (2012) NaCl salinity affects germination, growth, physiology, and biochemistry of bambara groundnut. Braz J Plant Physiol 24:151–160. https://doi.org/10.1590/S1677-04202012000300002

Arif Y, Singh P, Siddiqui H, Bajguz A, Hayat S (2020) Salinity induced physiological and biochemical changes in plants: an omic approach towards salt stress tolerance. Plant Physiol Biochem 156(7):64–77. https://doi.org/10.1016/j.plaphy.2020.08.042

Bangar P, Chaudhury A, Tiwari B, Kumar S, Kumari R, Bhat KV (2019) Morphophysiological and biochemical response of mungbean [*Vigna radiata* (L.) Wilczek] varieties at different developmental stages under drought stress. Turk J Biol 43(1):58–69. https://doi.org/10.3906/biy-1801-64

Baroowa B, Gogoi N, Farooq M (2016) Changes in physiological, biochemical and antioxidant enzyme activities of green gram (*Vigna radiata* L.) genotypes under drought. Acta Physiol Plant 38(9):219. https://doi.org/10.1007/s11738-016-2230-7

Bhardwaj R, Lone JK, Pandey R, Mondal N, Dhandapani R, Meena SK, Khan S, Gayacharan (2023) Insights into morphological and physio-biochemical adaptive responses in mungbean (*Vigna radiata* L.) under heat stress. Front Genet 14:1206451. https://doi.org/10.3389/fgene.2023.1206451

Breria CM, Hsieh CH, Yen TB, Yen JY, Noble TJ, Schafleitner R (2020) A SNP-based genome-wide association study to mine genetic loci associated to salinity tolerance in mungbean (*Vigna radiata* L.). Gene 11(7):759. https://doi.org/10.3390/genes11070759

Chand G, Nandwal AS, Kumar N, Devi S, Khajuria S (2018) Yield and physiological responses of mungbean *Vigna radiata* (L.) Wilczek genotypes to high temperature at reproductive stage. Legume Res-Int J 41(4):557–562

Chang Y, Peng L, Ji L, Wang S, Wang L, Wu J (2023) Genome-wise association study identified genomic regions associated with drought tolerance in mungbean (*Vigna radiata* (L.) R. Wilczek). Theor Appl Genet 136(3):40. https://doi.org/10.1007/s00122-023-04303-3

Chaudhary S, Priya M, Jha UC, Pratap A, HanumanthaRao B, Singh I, Prasad PV, Siddique KH, Nayyar H (2022) Approaches toward developing heat and drought tolerance in mungbean. Developing climate resilient grain and forage legumes 28:205–234. https://doi.org/10.1007/978-981-16-9848-4_10

Deeroum A, Thepphomwong K, Laosatit K, Somta P (2024) Inheritance of salt tolerance in wild mungbean (*Vigna radiata* var. sublobata). Agriculture and natural Resources 13:58(4)

Desai S, Mistry J, Shah F, Chandwani S, Amaresan N, Supriya NR (2023) Salt-tolerant bacteria enhance the growth of mung bean (*Vigna radiata* L.) and uptake of nutrients, and mobilize sodium ions under salt stress condition. Int J Phytoremediation 25(1):66–73. https://doi.org/10.1080/15226514.2022.2057419

Diatta AA, Abaye O, Battaglia ML, Leme JF, Seleiman M, Babur E, Thomason WE (2024) Mungbean [*Vigna radiata* (L.) Wilczek] and its potential for crop diversification and sustainable food production in Sub-Saharan Africa: a review. Technol Agron 4(1):e031. https://doi.org/10.48130/tia-0024-0030

Dutta P, Bandopadhyay P, Bera AK (2016) Identification of leaf based physiological markers for drought susceptibility during early seedling development of mungbean. Am J Plant Sci 7(14):1921–1936. https://doi.org/10.4236/ajps.2016.714176

Fatima A, Khan MJ, Awan HM, Akhtar MN, Bibi N, Sughra K, Khan MR, Ahmad R, Ibrahim M, Hussain J, Sadiq I (2020) Genome-wide identification and expression analysis of SnRK2 gene family in mungbean (*Vigna radiata*) in response to drought stress. Crop Pasture Sci 71(5):469–476

Hanif A, Wahid A (2018) Seed yield loss in mungbean is associated to heat stress induced oxidative damage and loss of photosynthetic capacity in proximal trifoliate leaf. Pak J Agric Sci 55(4):777–786. https://doi.org/10.21162/PAKJAS/18.7461

HanumanthaRao B, Nair RM, Nayyar H (2016) Salinity and high temperature tolerance in mungbean [*Vigna radiata* (L.) Wilczek] from a physiological perspective. Front Plant Sci 7:957. https://doi.org/10.3389/fpls.2016.00957

Hapsari RT, Trustinah T (2018) Salinity tolerance of mungbean genotypes at seedling stage. Biosaintifika 10(2):408–415. https://doi.org/10.15294/biosaintifika.v10i2.13999

Ikram S, Bhattarai S, Walsh KB (2022) Characterisation of selected mungbean genotypes for tolerance to waterlogging stress at pod filling stage. Agronomy 12(7):1663. https://doi.org/10.3390/agronomy12071663

Islam MT (2015) Effects of high temperature on photosynthesis and yield in mungbean. Bangladesh J Bot 44:451–454. https://doi.org/10.3329/bjb.v44i3.38553

Islam MR, Hamid A, Khaliq QA, Ahmed JU, Haque MM, Karim MA (2007) Genetic variability in flooding tolerance of mungbean (Vigna radiata L. Wilczek) genotypes. Euphytica 156(1):247–255. https://doi.org/10.1007/s10681-007-9372-z

Islam MR, Hamid A, Karim MA, Haque MM, Khaliq QA, Ahmed JU (2008) Gas exchanges and yield responses of mungbean (*Vigna radiata* L. Wilczek) genotypes differing in flooding tolerance. Acta Physiol Plant 30(5):697–707. https://doi.org/10.1007/s11738-008-0168-0

Islam M, Kamal M, Hossain M, Hossain J, Azam M, Akhter M, Hasan M, Al-Ashkar I, Almutairi K, Sabagh A, Rahman M (2022) Drought tolerance in mung bean is associated with the genotypic divergence, regulation of proline, photosynthetic pigment and water relation. Phyton 92(3):955. https://doi.org/10.32604/phyton.2023.025138

Jisha KC, Puthur JT (2016) Seed priming with BABA (β-amino butyric acid): a cost-effective method of abiotic stress tolerance in Vigna radiata (L.) Wilczek. Protoplasma 253:277–289. https://doi.org/10.1007/s00709-015-0804-7

Kandil AA, Arafa AA, Sharief AE, Ramadan AN (2012) Genotypic differences between two mungbean varieties in response to salt stress at seedling stage. Int J Agric Sci 4(7):278–283
Karim MA, Fukamachi H, Komori S, Ogawa K, Hidaka T (2003) Growth, yield and photosynthetic activity of Vigna radiata L. grown at different temperature and light levels. Plant Prod Sci 6(1):43–49. https://doi.org/10.1626/pps.6.43
Kaur S, Nayyar H (2015) Selenium fertilization to salt-stressed mungbean (*Vigna radiata* L. Wilczek) plants reduces sodium uptake, improves reproductive function, pod set and seed yield. Sci Hortic 197:304–317. https://doi.org/10.1016/j.scienta.2015.09.048
Kaur R, Bains TS, Bindumadhava H, Nayyar H (2015) Responses of mungbean (*Vigna radiata* L.) genotypes to heat stress: effects on reproductive biology, leaf function and yield traits. Sci Hortic 197:527–541. https://doi.org/10.1016/j.scienta.2015.10.015
Kobir MS, Ali MO, Hossain J, Alam MS, Rahman M, Ahammad KU, Hajong P, Paul S (2022) Salt stress tolerance and germination performance of mungbean genotypes under salt stress. Bangladesh Agron J 25(1):57–65. https://doi.org/10.3329/baj.v25i1.62832
Kumar BS, Prakash M, Narayanan S, Gokulakrishnan J (2012) Breeding for salinity tolerance in mungbean. APCBEE Proc 4:30–35. https://doi.org/10.1016/j.apcbee.2012.11.006
Kumar P, Pal M, Joshi R, Sairam RK (2013) Yield, growth and physiological responses of mung bean [*Vigna radiata* (L.) Wilczek] genotypes to waterlogging at vegetative stage. Physiol Mol Biol Plants 19(2):209–220. https://doi.org/10.1007/s12298-012-0153-3
Kumar A, Singh S, Gaurav AK, Srivastava S, Verma JP (2020) Plant growth-promoting bacteria: biological tools for the mitigation of salinity stress in plants. Front Microbiol 11:1216. https://doi.org/10.3389/fmicb.2020.01216
Kumari S, Vaishnav A, Jain S, Varma A, Choudhary DK (2016) Induced drought tolerance through wild and mutant bacterial strain pseudomonas simiae in mung bean (*Vigna radiata* L.). World J Microbiol Biotechnol 32(1):4. https://doi.org/10.1007/s11274-015-1974-3
Kyu KL, Taylor CM, Douglas CA, Malik AI, Colmer TD, Siddique KH, Erskine W (2024) Genetic diversity and candidate genes for transient waterlogging tolerance in mungbean at the germination and seedling stages. Front Plant Sci 15:1297096. https://doi.org/10.3389/fpls.2024.1297096
Labbo AM, Mehmood M, Akhtar MN, Khan MJ, Tariq A, Sadiq I (2018) Genome-wide identification of AP2/ERF transcription factors in mungbean (*Vigna radiata*) and expression profiling of the VrDREB subfamily under drought stress. Crop Pasture Sci 69(10):1009–1019. https://doi.org/10.1071/CP18180
Li S, Wang R, Jin H, Ding Y, Cai C (2019) Molecular characterization and expression profile analysis of heat shock transcription factors in mungbean. Front Genet 9:736. https://doi.org/10.3389/fgene.2018.00736
Liu C, Wu J, Wang L, Fan B, Cao Z, Su Q, Zhang Z, Wang Y, Tian J, Wang S (2017) Quantitative trait locus mapping under irrigated and drought treatments based on a novel genetic linkage map in mungbean (*Vigna radiata* L.). Theor Appl Genet 130(11):2375–2393. https://doi.org/10.1007/s00122-017-2965-6
Liu J, Xue C, Lin Y, Yan Q, Chen J, Wu R, Zhang X, Chen X, Yuan X (2022) Genetic analysis and identification of *VrFRO8*, a salt tolerance-related gene in mungbean. Gene 836:146658
Manasa R, Bindumadhava H, Nair RM, Prasad TG, Shankar AG (2017) Screening mungbean (Vigna radiata L.) lines for salinity tolerance using salinity induction response technique at seedling and physiological growth assay at whole plant level. No. AVRDC Staff Publication
Mankar GD, Wayase UR, Shelke DB, Nikam TD, Barmukh RB (2021) Morphological, physiological, and biochemical responses to NaCl-induced salt stress in mungbean (*Vigna radiata* L.) varieties. Notulae Scientia Biologicae 13(2):10936. https://doi.org/10.15835/nsb13210936
Mansoor S, Naqvi FN (2013) Effect of heat stress on lipid peroxidation and antioxidant enzymes in mungbean (Vigna radiata L) seedlings. Afr J Biotechnol 12(21):3196–3203
Manu B, Biradar R, Sabale PR, Kumar K, Aski MS, Mohite N, Shinde P, Kodandaram MH, Singh AK, Venkatesh MS, Mogali SC (2022) Genomic designing for abiotic stress tolerance in mungbean and urdbean. In: Genomic designing for abiotic stress resistant pulse crops. Springer International Publishing, Cham, pp 271–343. https://doi.org/10.1007/978-3-030-91039-6_8

Maryani Y, Dewi WS, Yunus A (2018) Study on osmoprotectant rhizobacteria to improve mung bean growth under drought stress. In: Paper presented at IOP conference series: earth and environmental science, Banda Aceh, Indonesia, 26–27 September 2018

Moradi A, Ahmadi AH, Hossain Zadeh A (2008) The effects of different timings and severity of drought stress on gas exchange parameters of mungbean. Desert 13(1):59–66. https://doi.org/10.22059/jdesert.2008.27176

Naher N, Alam AK (2010) Germination, growth and nodulation of mungbean (*Vigna radiata* L.) as affected by sodium chloride. Int J Sustain Crop Prod 5(2):8–11

Nair RM, Schafleitner R, Kenyon L, Srinivasan R, Easdown W, Ebert AW, Hanson P (2012) Genetic improvement of mungbean. SABRAO J Breed Genet 44:177–190

Narina SS, Pathak SC, Bhardwaj HL (2014) Chlorophyll fluorescence to evaluate pigeonpea breeding lines and mungbean for drought tolerance. J Agric Sci 6(11):238–246

Nivethitha T, Babu C, Jayamani P, Vijayalakshmi D (2024) Impact of drought stress on seedling vigor and physio-biochemical attributes of diverse mungbean genotypes for drought tolerance through in vitro and field evaluation. Euphytica 220(12):182. https://doi.org/10.1007/s10681-024-03438-8

Noble TJ, Williams B, Hoang TM, Das Bhowmik SS, Tan GZ, Mundree S (2020) Genomic approaches to abiotic stresses in mungbean. In: The Mungbean genome. Springer International Publishing, Cham, pp 169–179

Priya M, Pratap A, Sengupta D, Siddique KH, Singh NP, Jha U, Nayyar H (2020) Mungbean and high-temperature stress: responses and strategies to improve heat tolerance. In: Heat stress in food grain crops: plant breeding and omics research. Bentham Science Publishers, pp 144–170

Promila K, Kumar S (2000) *Vigna radiata* seed germination under salinity. Biol Plant 43(3):423–426

Rabie GH (2005) Influence of arbuscular mycorrhizal fungi and kinetin on the response mungbean plants to irrigation with seawater. Mycorrhiza 15:225–230. https://doi.org/10.1007/s00572-004-0345-y

Rahman MM, Habib MA, Sikdar MS, Shamsuzzoha M, Islam MS (2016) Evaluation of mungbean genotypes for salt tolerance at seedling stage and alleviation of saline stress by gypsum. Fundament Appl Agric 1(1):39–43

Raina SK, Govindasamy V, Kumar M, Singh AK, Rane J, Minhas PS (2016) Genetic variation in physiological responses of mungbeans (*Vigna radiata* (L.) Wilczek) to drought. Acta Physiol Plant 38(11):263. https://doi.org/10.1007/s11738-016-2280-x

Rajesh S, Kavya S, Navinraj S, Radhamani T, Anitha T, Ramesh SV (2025) Genome-wide identification and analysis of the SnRK2 gene family in cowpea (*Vigna unguiculata* (L.) Walp.) reveals potential stress-responsive roles. Plant Sci Today 12(1):1–6

Reardon ME, Qaderi MM (2017) Individual and interactive effects of temperature, carbon dioxide and abscisic acid on mungbean (*Vigna radiata*) plants. J Plant Interact 12(1):295–303. https://doi.org/10.1080/17429145.2017.1353654

Sadeghipour O (2019) Polyamines protect mung bean [Vigna radiata (L.) Wilczek] plants against drought stress. Biol Futura 70:71–78. https://doi.org/10.1556/019.70.2019.09

Sanghani JM, Mandaviya MK, Sanghani AO, Raval SS, Golakiya BA (2015) Analysis of genetic diversity among mung bean (*Vigna radiata* L.) genotypes through different biochemical markers. Indian J Agric Biochem 28(1):24–28

Sehrawat N, Bhat KV, Kaga A, Tomooka N, Yadav M, Jaiwal PK (2014) Development of new gene-specific markers associated with salt tolerance for mungbean (*Vigna radiata* L. Wilczek). Span J Agric Res 12(3):732–741

Sehrawat N, Yadav M, Bhat KV, Sairam RK, Jaiwal PK (2015) Effect of salinity stress on mungbean [*Vigna radiata* (L.) Wilczek] during consecutive summer and spring seasons. J Agric Sci, Belgrade 60(1):23–32. https://doi.org/10.2298/JAS1501023S

Sehrawat N, Yadav M, Sharma AK, Kumar V, Bhat KV (2019) Salt stress and mungbean [Vigna radiata (L.) Wilczek]: effects, physiological perspective and management practices for alleviating salinity. Arch Agron Soil Sci 65(9):1287–1301

Sharma L, Priya M, Bindumadhava H, Nair RM, Nayyar H (2016) Influence of high temperature stress on growth, phenology and yield performance of mungbean [*Vigna radiata* (L.) Wilczek]

under managed growth conditions. Sci Hortic 213:379–391. https://doi.org/10.1016/j.scienta.2016.10.033

Shukla PR, Skeg J, Buendia EC, Masson-Delmotte V, Pörtner HO, Roberts DC, Zhai P, Slade R, Connors S, Van Diemen S, Ferrat M (2019) Climate Change and Land: an IPCC special report on climate change, desertification, land degradation, sustainable land management, food security, and greenhouse gas fluxes in terrestrial ecosystems. https://doi.org/10.25561/76618

Singh DP, Singh BB (2011) Breeding for tolerance to abiotic stresses in mungbean. J Food Legumes 24(2):83–90

Singh CM, Kumar M, Pratap A, Tripathi A, Singh S, Mishra A, Kumar H, Nair RM, Singh NP (2022) Genome-wide analysis of late embryogenesis abundant protein gene family in vigna species and expression of VrLEA encoding genes in vigna glabrescens reveal its role in heat tolerance. Front Plant Sci 13:843107. https://doi.org/10.3389/fpls.2022.843107

Xu N, Chen B, Cheng Y, Su Y, Song M, Guo R, Wang M, Deng K, Lan T, Bao S, Wang G (2023) Integration of GWAS and RNA-seq analysis to identify SNPs and candidate genes associated with alkali stress tolerance at the germination stage in mungbean. Genes 14(6):1294. https://doi.org/10.3390/genes14061294

Zandalinas SI, Mittler R, Balfagón D, Arbona V, Gómez-Cadenas A (2018) Plant adaptations to the combination of drought and high temperatures. Physiol Plant 162(1):2–12

Gene Cloning and Function

Wang Lixia

Abstract

As a traditional crop in east or southeast Asia, the genetic study on mungbean is much lagged. In the 1930s, Bose discovered that the color of young pods and flower was controlled by the same pair of genes and that seed coat gloss is a recessive trait controlled by a single gene. In recent years, the demand for mungbean has increased. This has led to greater focus on its genetic research. Studies on genetic effects and heritability of agronomic traits, such as yield and disease resistance, have been widely conducted. These efforts have provided a strong base for molecular genetics research in mungbean.

Keywords

Gene cloning · Gene mapping · Inheritance traits · Mungbean

1 Introduction

The genetic study on mungbean is much lagged owing to the small scale of production and fairly narrow distribution in the last century. However, there have been great progresses in the recent decades, because the promising future of mungbean, such as its excellent nutrition, short growth period and Nitrogen fixation. Here we reviewed the genetic study of mungbean including the gene clone and function investigation, to provide reference to those engaging in mung bean biological research.

W. Lixia (✉)
Key Laboratory Grain Crop Genetic Resources Evaluation and Utilization, Ministry of Agriculture and Rural Affairs, Institute of Crop Sciences, Chinese Academy of Agricultural Sciences (ICS-CAAS), Beijing, China
e-mail: wanglixia03@caas.cn

S. Paramasivam et al. (eds.), *Mungbean*,
https://doi.org/10.1007/978-981-95-7637-1_8

2 Inheritance of Traits in Mungbean

The most studied traits were yield-related characteristics, which could provide important information for breeding and production. A study involving eight mungbean genotypes and their hybrids examined five yield-related traits. The results showed that dominance effects played a major role in controlling these traits. Epistatic interactions were also observed for most traits. These findings support the use of diverse parents in breeding programs. To avoid loss of superior lines, selection is best carried out in advanced generations (Aher et al. 1999). Generation mean analysis of quantitative traits of mungbean hybrid combinations showed that there were additive effects on pod number per plant and 100-seed weight, while there were dominant effects on pod number per plant and yield per plant (Kute et al. 2003). Both additive and dominance gene effects were observed for pod number per plant, 100-seed weight, yield per plant, and biological yield. Additive effects were more important for these traits. Therefore, effective improvement can be achieved through selection in early generations, which reduces workload and improves breeding efficiency (Khattak et al. 2002; Jayaprada et al. 2005). Other studies have shown additive, dominant, and epistatic effects of seed yield-related gene, which are consistent with Singh et al. (2006). As an important trait for mungbean, seed color was reported to be controlled by a single gene, and green was dominant to yellow, while black was dominant to green (Wang et al. 2014).

Bruchid is one of the most serious pests of mungbean and has attracted the attention of many scientists. Yield losses of 30–50% occur in normal years. In severe infestations, complete yield loss may occur. The weight and quality of mungbean damaged by weevil are reduced, which seriously affects the commodity quality and market price of mungbean (Talekar 1988). The control of weevils using fumigation and chemical insecticides is costly and often causes environmental pollution. Therefore, breeding weevil-resistant mungbean varieties has become a major breeding goal (Srinives 1995). Most identified resistant germplasm, however, shows small seed size, hard grains, and poor agronomic performance. As a result, their use in breeding programs is limited (Tomooka et al. 1992; Watanasit and Pichitporn 1995). Studying the inheritance of weevil resistance in wild mungbean can improve the effective use of these genes in breeding. TC1966, a wild mungbean accession from Madagascar, shows complete resistance to both mungbean weevil and four-lined weevil. Kitamura et al.'s analysis showed that the ratio of TC1966 to cultivar F2 was consistent with the phenotypic segregation of 3:1, the ratio of F3 genotype was consistent with the phenotypic segregation of 1:2:1, and the resistance gene of TC1966 was controlled by a single dominant gene (Kitamura et al. 1988), which was further verified by Cheng et al. (2005). Lin et al. showed that more than 95% of worm eggs could not hatch on the surface of VC6089A grain, Vrd1 extracted from VC6089A grain was toxic to weevil, and the eggs could not develop normally on the surface of artificial seeds containing 0.2% Vrd1, suggesting that VRD1 may be a major component of resistance to bruchid (Lin et al. 2005). However, further study identified some individuals with high VRD1 content but sensitive to weevil (Kaga and Ishimoto 1998). Therefore, VRD1 gene might only be closely linked to the resistance gene to bruchid. These results will provide useful information for the utilization of resistance genes in TC1966 in breeding, such as marker-assisted identification and target gene cloning. Wild Mungbean ACC41, from Australia, also has a broad spectrum of insect resistance, and studies have shown that the insect resistance of ACC41 is controlled by a single dominant gene (Lambrides and Imrie 2000). In addition to the two wild species mentioned above, AVRDC also identified two mungbean cultivars V2709 and V2802 which were resistant to bruchid, but their resistance to bruchid was moderate. Somta et al. showed that the resistance to legume in V2709 and V2802 was controlled by a pair of major genes, but there was the effect of modified genes, and the resistance genes between them might be alleles (Somta et al. 2007). V6089, V2709, and V2802, which were bred by TC1966, were crossed artificially in the field, and the segregation ratio of resistant plants in the progenies showed that the resistance gene of TC1966 was not allelic to V2709 and V2802 (unpublished), which provided important information for the breeding of new varieties by pyramiding resistance genes. In addition to weevil, leaf spot, powdery mildew, and mosaic virus diseases are also harmful to mungbean production. At present, there are also many reports on the inheritance of mungbean disease resistance. Two lines resistant to powdery mildew was crossed and two lines susceptible to powdery mildew. The analysis of the mean value of different generations showed that there were significant additive and dominant effects of genes for resistance to powdery mildew in mungbean, and the segregation ratio of resistance to powdery mildew in BC1 and F_2 generations of backcross showed that the resistance of the four combinations was controlled by a single dominant gene (Chaitieng et al. 2022). However, the F1 hybrids of mungbean showed moderate resistance to the disease, and the segregation ratio of resistance and susceptibility in F_2 and F_3 generations suggested that the resistance to powdery mildew was controlled by three recessive genes (Kute et al. 2003). These results indicated that the genetic mechanism of resistance was different among different resistance sources, and the resistance genes were not allelic. The results showed that the resistance of mungbean to mosaic virus was controlled by a single recessive gene, but the existence of modified genes might lead to a difference in resistance in the progeny population; however, resistance to mosaic virus is controlled by two pairs of recessive genes in black gram, a relative species of mungbean losely related species, black gibbon, probably due to differences between hosts (Khattak et al. 2000; Ammavasai et al. 2004; Sirohi et al. 2002). The existence of the above-mentioned multi-resistance genes is of great significance for gene pyramiding breeding. Miyagi et al. (2004) developed PCR-based markers that closely linked to a major locus conditioning bruchid resistance using bacterial artificial chromosome libraries.

3 Gene Mapping in Mungbean

The early mappings were not so fine, owning to the lack of numerous markers. By constructing a genetic linkage map, Young et al. found an RFLP marker only 3.6 cM from the resistance gene in wild Mungbean TC1966 in Linkage Group 8 (Young et al. 1992). Six markers closely linked to the resistance gene of TC1966 were found, and the resistance traits of the backcross progeny population were identified, and a single susceptible plant was found to detect mungbean acid (Kaga and Ishimoto 1998), thus denying the view of Sugawara et al. that mungbean acid is one of the main factors of resistance to weevil (Sugawara et al. 1996). However, it was also suggested that glycine max may be closely linked to the resistance gene, so glycine max may be useful in accelerating the identification of resistance traits and map-based cloning of resistance genes. Based on the identification under different enviroments, the resistance to bruchid was stable in the wild mungbean, ACC41, and was deemed to be an ideal target for further fine mapping (Mei et al. 2009). All these studies laid a foundation for fine mapping, cloning, and marker-assisted selection breeding of target genes for bruchid resistance. In addition to mapping and QTL analysis for resistance to bruchid, a genetic map using 147 recombinant inbred lines (RFLP) derived from Indian Mungbean ATF3640 with high resistance to powdery mildew and susceptible cultivars, was constructed. The total length of the map was 350 cM and it contained 52 marker loci. Based on the map, a major QTL was developed that could explain 86% of the genetic variation, and the distance between the marker and the powdery mildew resistance gene was only 1.3 cM (Humphry et al. 2003). Hard seedness is an serious problem for the commodity quality of Mungbean, including boiling and processing, but it is much beneficial for production in the regions with abundant rainfall by avoiding seed rot or sprout before harvesting in the years with abundant rainfall, therefore, breeding Mungbean varieties with higher rate of hard seed is also one of the objectives of Mungbean breeding in regions with more rainfall. The segregation ratios of hard-seeded grains in the offspring of different combinations suggested that hardseedness was controlled by a single dominant gene, indicating that it is possible to develop varieties with low or high level of hard seedness percentage are small grains, it is generally believed that it is very difficult to breed new varieties of Mungbean with large grains and high hard seed rate. However, studies on QTLS for grain size and hard seed setting showed that some of the overlapping qtls exist stably in different environments, so it can be considered that it is possible to breed mungbean varieties with large and hard seed setting (Humphry et al. 2005). Singh et al. (2021) identified loci with agronomic traits using SSR markers. Rohilla et al. (2022) identified loci related to resistance toward yellow mosaic disease (YMD) using 80 cultivars by 164 genome-wide SSRs. Kyu et al. (2024) investigated the diversity of transient water logging tolerance in a mini-core germplasm collection of mungbean and identified candidate genes for adaptive traits of interest using genome-wide association studies (GWAS) at two critical stages of growth: germination and seedling stage. Azeem et al. (2021) identified 39 genes encoding putative K+ transport-related genes in mungbean, further analysis suggested that gene duplication is a major route of expansion for this family in mungbean, and these genes might play roles in abiotic stress tolerance. With the development of molecular markers in mungbean, numerous studies using high-through output sequencing were published. Based on GWAS, Wu et al. (2020) identified a set of SNP association with seed mineral concentration, and SokolkovaA et al. (2020) identified several loci significantly associated with agronomical traits. While Sinha et al. (2023) revealed that a total of 20 significant SNPs were shared by both GLM and Blink models associated with grain micronutrient and anti-nutritional factor traits. Reddy et al. (2020) identified SNPs related with phosphorus use efficiency. Based on genomic and transcriptomic analysis, Liu et al. (2016) identified 91 differentially expressed genes (DEGs), which were classified into 17 major and 74 minor bruchid-resistance-associated genes. Lin et al. (2023a, b) mapped the QTL conferring resistance to calcareous soil using a RIL population and identified VrYSL3 that was significantly upregulated in calcareous soil.

4 Gene Cloning and Function

Both the inheritance and gene mapping laid an important foundation for further study, especially for gene cloning. However, up to now, only limited genes were cloned, either by fine mapping or homologous sequences. At present, although the gene cloning and expression analysis of mungbean have started, the research content is still scattered and systematically poor. For example, cloning and analysis of the cDNA sequence of mungbean acid VRD1 showed that the expressed product of the recombinant gene could inhibit the growth of fungi and the development of weevil larvae, For example, cloning and analysis of the cDNA sequence of mungbean acid VRD1 showed that the expressed product could inhibit the growth of fungi and the development of weevil larvae, as well as affecting protein synthesis in vitro (Chen et al. 2004a). Miao et al. also performed sequence analysis and vector construction of mungbean defensin gene, but did not identify the expression products of target genes (Liao et al. 2006). Chen et al. isolated the cDNA of Mungbean HSC70 and performed RNA and protein expression level analysis, and the results showed that this gene mainly plays a role in the normal growth and development process, which belongs to the constitutive expression gene (Chen et al. 2004b). Isolation of a hormone-induced protease gene from mungbean and expression analysis of the transgene suggest that this gene is specifically induced by auxin and may play an important role in protein phosphorylation in hormone signaling (Kwon et al. 2004). All of the above methods were obtained by genomic PCR with primers designed with known protein sequences, and there was no way to deal with the gene whose expression product was not clear or whose sequence was completely unknown. Therefore, map-based cloning and large-scale library screening have great application potential in mungbean gene cloning. Lin et al. (2023a, b) identified the gene associated with the pod shattering resistance trait, VrMYB26a, and the stem determinacy trait, VrDet1, by using whole-genome resequencing; they also found that two ancient haplotypes of VrDet1 have lower gene expression and exhibited intermediate frequencies. Based on transcriptome analysis, Jiao et al. (2019) found that a set of developmental regulators were altered in rachis and fused leaflets mutants compared to the wild-type plants. Yin et al. (2024) analyzed the genomic characteristics and expression patterns of the BBX gene family under ABA, PEG, and NaCl treatments. They proposed that VrBBX genes may play a role in response to environmental stress and VrBBX5, VrBBX10, and VrBBX12 are important candidate genes for plant stress response. While Kang et al. (2017) analyzed the genome-wide DNA methylation profile in mungbean and found that genes differentially expressed between different cultivars showed distinct DNA methylation patterns. Bharadwaj et al. (2019) isolated an identified potential stress responsive candidate gene SKP1, and the upregulation in relative expression level of this gene in different plant parts under drought stress confirmed its utility in drought tolerant breeding.

5 Conclusion

With the development of DNA markers and sequencing technology, significant progress has been made in the cloning of mungbean genes, especially in the research of disease resistance and stress tolerance genes. However, the technical system of mungbean genetic transformation has not yet been successfully constructed, which will be an obstacle for the verification of gene function. Therefore, it is necessary to further strengthen the research and development of mungbean genetic transformation technology.

References

Aher RP, Dahat DV, Sonowane VP (1999) Combining ability studies in Mungbean [*Vigna* radiata (L.) Wilczek]. Crop Res (Hisar) 18(2):256–260

Ammavasai S, Phogat DS, SolankiI S (2004) Inheritance of resistance to Mungbean yellow mosaic virus (MYMV) in green gram [*Vigna radiata* (L.)Wilczek]. Indian J Genet Plant Breed 64:146

Azeem F, Ijaz U, Ali MA, Hussain S, Zubair M, Manzoor H, Abid M, Zameer R, Kim DS, Golokhvast KS, Chung G, Sun S, Nawaz MA (2021) Genome-wide identification and expression profiling of potassium transport-related genes in *Vigna radiata* under abiotic stresses. Plants (Basel) 11(1):2

Bharadwaj N, Barthakur S, Biswas AD, Kumar Das M, Kour M, Ramteke A, Gogoi N (2019) Transcript expression profiling in two contrasting cultivars and molecular cloning of a SKP-1 like gene, a component of SCF-ubiquitin proteasome system from Mungbean *Vigna radiata* L. Sci Rep 9(1):8103. https://doi.org/10.1038/s41598-019-44034-4

Chaitieng B, Laosuwan P, Wongkaew S (2022) Inheritance of powdery mildew resistance in Mungbean [*Vigna radiata* (L.)Wilczek]. Thail J Agric Sci 36:73–78

Chen JJ, Chen GH, Hsu HC, Li SS, Chen CS (2004a) Cloning and functional expression of a MungbeanDefensin VrD1 in Pichiapastoris. J Agric Food Chem 52:2256–2261. https://doi.org/10.1021/jf030662i

Chen YJ, Wu MF, Yu YH, Tam MF, Lin TY (2004b) Developmental expression of three Mungbean Hsc70s and substrate-binding specificity of the encoded proteins. Plant Cell Physiol 45:1603–1614. https://doi.org/10.1093/pcp/pch182

Cheng XZ, Wang SH, Wu SY, Zhou JH, Wang SM, Yang YD (2005) Tagging and utilization of bruchid resistance gene using PCR markers in Mungbean. Sci Agric Sin 38:1534–1539. https://doi.org/10.1093/pcp/pch182

Humphry ME, Magner T, McIntyre CL, Aitken EA, Liu CJ (2003) Identification of a major locus conferring resistance to powdery mildew (*Erysiphe polygoni* DC) in Mungbean (*Vigna radiata* (L.)Wilczek) by QTL analysis. Genome 46:738–744. https://doi.org/10.1139/g03-057

Humphry ME, Lambrides CJ, Chapman SC, Aitken EAB, Imrie BC, Lawn RJ, McIntyre CL, Liu CJ (2005) Relationships between hard-seededness and seed weight in Mungbean

(*Vigna radiata*) assessed by QTL analysis. Plant Breed 124:292–298. https://doi.org/10.1111/j.1439-0523.2005.01084.x

Jayaprada M, Reddy KR, Sekhar MR, Reddy GL (2005) Genetic analysis of quantitative characters in Mungbean [*Vigna radiata* (L.)Wilczek]. Legum Res 28:210–212

Jiao K, Li X, Guo Y, Guan Y, Guo W, Luo D, Hu Z, Shen Z (2019) Regulation of compound leaf development in Mungbean (*Vigna radiata* L.) by cup-shaped cotyledon/no apical meristem (CUC/NAM) gene. Planta 249(3):765–774. https://doi.org/10.1007/s00425-018-3038-z

Kaga A, Ishimoto M (1998) Genetic localization of a bruchid resistance gene and its relationship toinsecticidalcyclopeptide alkaloids, the *Vignatic* acids, in Mungbean (*Vigna radiata* (L.) Wilczek). Mol Gen Genet 258:378–384. https://doi.org/10.1007/s004380050744

Kang YJ, Bae A, Shim S, Lee T, Lee J, Satyawan D, Kim MY, Lee SH (2017) Genome-wide DNA methylation profile in Mungbean. Sci Rep 7:40503. https://doi.org/10.1038/srep40503

Khattak GSS, Haq MA, Ashraf M (2000) Genetics of Mungbean yellow mosaic virus (MYMV) in Mungbean (*Vigna radiata* (L.) Wilczek) wilczek. J Genet Breed 54:237–243

Khattak GS, Ashraf M, Haq MA, McNeilly T, Rha ES (2002) Genetic basis of plant height and its degree of indetermination in Mungbean (*Vigna radiata* (L.) Wilczek). Hereditas 137(1):52–56. https://doi.org/10.1034/j.1601-5223.2002.1370107.x

Kitamura K, Ishimoto M, Sawa M (1988) Inheritance of resistance to infestation with azuki bean weevil in *Vigna sublobata* and successful incorporation to *V. radiata*. Jpn J Breed 38:459–464. https://doi.org/10.1270/jsbbs1951.38.459

Kute NS, Deshmukh RB, Patil JV (2003) Inheritance of resistance to powdery mildew (*Erysiphe polygoni* D.C.) in Mungbean (*Vigna radiata* (L.) Wilczek). Legum Res 26:24–27

Kwon C, Yun HS, Kaufman PB, Kim SK, Kim TW, Kang BG, Chang SC (2004) Identification and characterization of an auxin-inducible protein kinase, VrCRK1, from Mungbean. Mol Cells 18:346–352. https://doi.org/10.1016/S1016-8478(23)13123-2

Kyu KL, Taylor CM, Douglas CA, Malik AI, Colmer TD, Siddique KHM, Erskine W (2024) Genetic diversity and candidate genes for transient waterlogging tolerance in Mungbean at the germination and seedling stages. Front Plant Sci 15:1297096. https://doi.org/10.3389/fpls.2024.1297096

Lambrides CJ, Imrie BC (2000) Susceptibility of Mungbean Varieties to the Bruchid Species *Callosobruchus Maculatus* (F.), C. Phaseoli (Gyll.), *C. chinensis* (L.), and Acanthoscelides Obtectus (Say.) (Coleoptera:Chrysomelidae). Aust J Agric Res 51:85–89. https://doi.org/10.1071/AR99051

Liao JK, Sun L, Peng XM, Zhu JB (2006) Cloning and sequencing of *Vigna* radiation defensin gene and construction of plant expression vector. J Shihezi Univ 24:112–115. (in Chinese)

Lin C, Chen CS, Horng SB (2005) Characterization of resistance to *Callosobruchus maculatus* (Coleoptera: Bruchidae) in Mungbean variety VC6089A and its resistance-associated protein VrD1. J Econ Entomol 98(4):1369–1373. https://doi.org/10.1603/0022-0493-98.4.1369

Lin Y, Amkul K, Laosatit K, Liu J, Yimram T, Chen J, Yuan X, Chen X, Somta P (2023a) Fine mapping of QTL conferring resistance to calcareous soil in Mungbean reveals VrYSL3 as candidate gene for the resistance. Plant Sci 332:111698. https://doi.org/10.1016/j.plantsci.2023.111698

Lin YP, Chen HW, Yeh PM, Anand SS, Lin J, Li J, Noble T, Nair R, Schafleitner R, Samsononova M, Bishop-von-Wettberg E, Nuzhdin S, Ting CT, Lawn RJ, Lee CR (2023b) Demographic history and distinct selection signatures of two domestication genes in Mungbean. Plant Physiol 193(2):1197–1212. https://doi.org/10.1093/plphys/kiad356

Liu C, Fan B, Cao Z, Su Q, Wang Y, Zhang Z, Tian J (2016) A deep sequencing analysis of transcriptomes and the development of EST-SSR markers in Mungbean (*Vigna radiata*). J Genet 95:527–535. https://doi.org/10.1007/s12041-016-0663-9

Mei L, Cheng XZ, Wang SH, Wang LX, Liu CY, Sun L, Xu N, Humphry ME, Lambrides CJ, Li HB, Liu CJ (2009) Relationship between bruchid resistance and seed mass in Mungbean based on QTL analysis. Genome 52(7):589–596. https://doi.org/10.1139/G09-031

Miyagi M, Humphry M, Ma ZY, Lambrides CJ, Bateson M, Liu CJ (2004) Construction of bacterial artificial chromosome libraries and theirapplication in developing PCR-based markers closely linked to a major locus conditioning bruchid resistance in Mungbean. Theor Appl Genet 110:151–156. https://doi.org/10.1007/s00122-004-1821-7

Reddy VRP, Das S, Dikshit HK, Mishra GP, Aski M, Meena SK, Singh A, Pandey R, Singh MP, Tripathi K, Gore PG, Priti, Bhagat TK, Kumar S, Nair R, Sharma TR (2020) Genome-wide association analysis for phosphorus use efficiency traits in Mungbean (*Vigna radiata* L. Wilczek) using genotyping by sequencing approach. Front Plant Sci 2020(11):537766. https://doi.org/10.3389/fpls.2020.537766

Rohilla V, Yadav RK, Poonia A, Sheoran R, Kumari G, Shanmugavadivel PS, Pratap A (2022) Association mapping for yield attributing traits and yellow mosaic disease resistance in Mungbean [*Vigna radiata* (L.) Wilczek]. Front Plant Sci 12:749439. https://doi.org/10.3389/fpls.2021.749439

Singh I, Gill MS, Bains TS (2006) Generation mean analysis for yield attributing traits in Mungbean [*Vigna radiata* (L.) Wilczek]. Indian J Genet Plant Breed 66:47–48

Singh CM, Prajapati U, Gupta S, Pratap A (2021) Microsatellite-based association mapping for agronomic traits in Mungbean (*Vigna radiata* L. Wilczek). J Genet 100:87

Sinha MK, Aski MS, Mishra GP, Kumar MBA, Yadav PS, Tokas JP, Gupta S, Pratap A, Kumar S, Nair RM, Schafleitner R, Dikshit HK (2023) Genome wide association analysis for grain micronutrients and anti-nutritional traits in Mungbean [*Vigna radiata* (L.) R. Wilczek] using SNP markers. Front Nutr 10:1099004. https://doi.org/10.3389/fnut.2023.1099004

Sirohi A, Singh A, Bishnoi SS (2002) Interaction of Mungbean yellow mosaic virus resistance in black gram (*Vigna mungo*). Crop Improv 29:173–176

Sokolkova A, Burlyaeva M, Valiannikova T, Vishnyakova M, Schafleitner R, Lee C, Ting C, Nair RM, Nuzhdin S, Samsonova M, Wettberg E (2020) Genome-wide association study in accessions of the mini-core collection of Mungbean(*Vigna radiata*) from the World Vegetable Gene Bank (Taiwan). BMC Plant Biol 20(Suppl 1):363

Somta P, Ammaranan C, Ooi PAC, Srinives P (2007) Inheritance of seed resistance to bruchids in cultivated Mungbean (*Vigna radiata*, L. Wilczek). Euphytica 155:47–55

Srinives P (1995) Mungbean breeding: past, present and future. In: Srinives P, Kitbamroong C, Miyazaki S (eds) Mungbeangermplasm: collection, evaluation and utilization for breeding program. Japan International Research Center for Agricultural Sciences, Tsukuba, p 73

Sugawara F, Ishimoto M, Levan N, Koshino H, Uzawa J, Yoshida S (1996) Insecticidal peptide from Mungbean—a resistant factor against infestation Azuki bean weevil. J Agric Food Chem 44(10):3360–3364. https://doi.org/10.1021/jf960166c

Talekar NS (1988) Biology, damage and control of bruchid pests of Mungbean. In: Shanmugasundaram S, McLean BT (eds) Mungbean: proceedings of the second international symposium. AVRDC, Tainan, pp 329–342

Tomooka N, Lairungruang C, Nakeeraks P, Egawa Y, Thavarasook C (1992) Development of bruchid resistant Mungbean using wild Mungbeangermplasm in Thailand. Plant Breed 109:60–66. https://doi.org/10.1111/j.1439-0523.1992.tb00151.x

Wang L, Cheng XW et al (2014) Phenotypic variation and ecological adaptability evaluation of Chinese Mungbean core germplasm resources under different environments. Acta Agron Sin 40(04):739–744

Watanasit A, Pichitporn S (1995) Improvement of Mungbean for resistance to bruchids. In: Srinives P, Kitbamroong C, Miyazaki S (eds) Mungbeangermplasm: collection, evaluation and utilization for breeding program. Japan International Research Center for Agricultural Sciences, Tsukuba, p 67

Wu X, Islam ASMF, Limpot N, Mackasmiel L, Mierzwa J, Cortés AJ, Blair MW (2020) Genome-wide SNP identification and association mapping for seed mineral concentration in Mungbean (*Vigna radiata* L.). Front Genet 11:656. https://doi.org/10.3389/fgene.2020.00656

Yin L, Wu R, An R, Feng Y, Qiu Y, Zhang M (2024) Genome-wide identification, molecular evolution and expression analysis of the B-box gene family in Mungbean (*Vigna radiata* L.). BMC Plant Biol 24(1):532. https://doi.org/10.1186/s12870-024-05236-9

Young ND, Kumar L, Menancio-Hautea D, Danesh D, Talekar NS, Shanmugasundaram S, Kim DH (1992) RFLP mapping of a major bruchid resistance gene in Mungbean (*Vigna radiata* L. Wilczek). Theor Appl Genet 84:839–844. https://doi.org/10.1007/BF00227394

In Vitro Regeneration and Genetic Transformation of Mungbean

Sivakumar Paramasivam, Jothi Kanmani Bharathi, Sanyasi Selvamani, and Shen Wang

Abstract

Mungbean (*Vigna radiata* (L.) R. Wilczek) is a significant legume crop globally, specifically in Asia, where it is used as the main source of protein. It has a short growth cycle, it fixes nitrogen, and it is nutritious, all of which support its place in sustainable agriculture and food security. Conventional methods of breeding and propagation have limitations because mungbean is seasonally dependent, lacks genetic diversity, and has low transformation efficiency. This review comprehensively examines the progress in in vitro regeneration and genetic transformation in mungbean and aims to address and overcome the existing limitations. Direct and indirect organogenesis has been achieved using various explants (cotyledonary nodes, shoot tips, hypocotyls, and embryonic axes) and the formulations media supplemented with cytokinins such as BAP and TDZ provide high-frequency shoot regeneration. Another reliable alternative explored is somatic embryogenesis, a type of embryo development that has shown promising results in certain genotypes. Regeneration efficiency also depends on the age of explants, hormonal balance, and growth conditions in culture which are highlighted in the studies too. Despite the important advances having been made, it is necessary to perform additional research in order to reduce the genotype-independent protocols, optimize liquid culture systems, and also minimize the

S. Paramasivam (✉)
Dr.M.S.Swaminathan Agricultural College and Research Institute, Tamil Nadu Agricultural University, Thanjavur, Tamil Nadu, India
e-mail: sivakumar.p@tnau.ac.in

J. Kanmani Bharathi · S. Selvamani
Faulty of Agriculture, Annamalai University, Annamalainagar, Tamil Nadu, India

S. Wang
Institute of Cereal and Oil Crops, Hebei Academy of Agricultural and Forestry Sciences/ Hebei Laboratory of Crop Genetic and Breeding, Shijiazhuang, Hebei, China

S. Paramasivam et al. (eds.), *Mungbean*,
https://doi.org/10.1007/978-981-95-7637-1_9

bottlenecks associated with transformation. Such developments will eventually enable massive propagation, functional genomics, and crop enhancement of *V. radiata* that will enable its further use to achieve sustainable agriculture.

Keywords

Mungbean · Tissue culture · Organogenesis · Somatic embryogenesis · Transformation

1 Introduction

Vigna radiata (L.) R. Wilczek, also referred to as green gram or mungbean, is one of the most economically significant legume crops globally (Baza et al. 2022). Millions of people in Asia rely on this short-duration pulse crop as their main supply of high-quality protein, making it an essential part of global food security strategies (Dahiya et al. 2015).*Vigna radiata* is a member of the Fabaceae family, specifically the subfamily Faboideae, which includes the genus *Vigna* and the subgenus Ceratotropis. This diploid species (2n = 2x = 22 chromosomes) exhibits an annual growth pattern, attaining heights of 15–125 cm, with robust root systems and trifoliate leaves. The plant yields pale yellow flowers in clusters of 12–15, which develop into cylindrical pods holding 10–15 seeds that exhibit colors ranging from green to yellow, brown, or mottled patterns (Swamy 2023). Mungbean offers significant nutritional benefits, with protein levels between 14.6 and 33.0 g per 100 g and iron quantities ranging from 5.9 to 7.6 mg per 100 g. This bean functions as a readily digestible protein source, especially vital for combating malnutrition in underdeveloped nations where it complements cereal-based diets (Dahiya et al. 2015). The crop's capacity to assimilate atmospheric nitrogen via symbiotic associations with rhizobia significantly augments its agricultural worth by enhancing soil fertility and diminishing fertilizer dependence (Pataczek et al. 2018).

Since the need for quick, disease-free plant multiplication is growing and agricultural needs are expanding, it is more important than ever to establish effective in vitro regeneration techniques for *V. radiata*. Plant regeneration has several benefits, but it's especially pertinent to initiatives aimed at improving legumes. These include facilitating genetic transformation, conserving endangered species, producing huge quantities of plants in a smaller amount of land, and producing genetically homogeneous, disease-free plants quickly (Bhatia et al. 2015). For crops such as *V. radiata*, where traditional propagation techniques could be constrained by needs for genetic consistency and seasonality, micropropagation presents previously unheard-of possibilities for mass production and crop improvement (Loberant and Altman 2010).

Over the past 20 years, research on *V. radiata* plant regeneration has advanced considerably, yet there are still issues. Several studies have shown that different explant forms, such as cotyledonary nodes, shoot tips, and cotyledon-attached decapitated embryos, can successfully regenerate new shoots (Bhajan et al. 2024).

The most successful procedures usually use Murashige and Skoog (MS) medium enhanced with particular cytokinin concentrations, including kinetin (Kn) and 6-benzylaminopurine (BAP) (Yadav et al. 2010). Vigna radiata remains difficult to use in tissue culture applications due to its recalcitrant nature, despite these advancements (Nivya and Shah 2023). Compared with other crop families, legumes generally exhibit lower transformation efficiencies; for example, V. radiata shows transformation rates of only 1.49–4.2%. Compared to other crop families, legumes often have poorer transformation efficiency; *V. radiata*, for example, only has transformation rates of 1.49–4.2%. This resistance is explained by species-specific defensive mechanisms that reduce the effectiveness of transformation, poor integration of foreign DNA, and challenges with regeneration under selection pressure (Hesami et al. 2024).

Recent breakthroughs in *V. radiata* regeneration research have focused on addressing traditional restrictions through new methodologies. Advanced techniques using nuclear technologies and tissue culture combinations have shown promise for boosting mutation frequency and enabling broader genetic variety for selection. The integration of molecular markers with micropropagation procedures has the potential to cut breeding program schedules dramatically (Yousif 2020). Somatic embryogenesis has become a viable substitute for conventional organogenesis methods. Using somatic embryogenesis and primary leaf explants, (Sivakumar et al. 2010) created an effective regeneration strategy that achieved a 60% regeneration frequency. This method successfully transformed somatic embryos into plantlets, potentially resolving the rooted issues that frequently arise in tissue culture of *V. radiata.* Looking ahead, a number of important areas need concentrated research effort. These include creating genotype-independent methods to get around variety-specific restrictions, integrating automated systems to save production costs, and optimizing liquid culture techniques for large-scale production. Achieving commercially effective *V. radiata* plant regeneration systems may be possible with the combination of sophisticated bioreactor technology and improved culture media compositions.

2 In Vitro Regeneration of Mungbean

Tropical grain legumes particularly mungbean exhibit diverse responses compared to other plant species when cultured in vitro. High frequency in vitro regeneration protocol is limited in most of the tropical grain legumes particularly mungbean. This opens up the possibility to standardize the high frequency in vitro regeneration protocol in mungbean either through organogenesis or somatic embryogenesis.

2.1 Plant Regeneration Through Organogenesis

The successful development of in vitro regeneration protocol in mungbean is mainly dependent on the nature, type, size, and age of the explants and the type of medium compositions used for culture growth. (Avenido and Hautea 1990; Gulati and Jaiwal 1990, 1992, 1994; Chandra and Pal 1995; Girija et al. 2000; Tivarekar and Eapen 2001; Jaiwal et al. 2001; Amutha et al. 2003, 2006; Devi et al. 2004; Mundhara and Rashid 2006; Sivakumar et al. 2010; Vijayan et al. 2006; Sahoo et al. 2016; Tripathi et al. 2022; Rathore et al. 2023).

Mathews and Rao (1984) developed a technique for enhancing regeneration efficiency by culturing cotyledons excised from imbibed seeds, which directly regenerated into whole plants. Notably, they observed an inverse relationship between the duration of seed imbibition and regeneration potential, suggesting that physiological status of the explant prior to culture plays a critical role. Taken together, these studies consistently underscore the suitability of cotyledons as responsive explants and emphasize the synergistic role of explant type, developmental stage, and hormonal composition in establishing efficient in vitro regeneration systems for mungbean. Mathews (1987) demonstrated that various seedling explants such as cotyledons, cotyledonary nodes, shoot tips, and roots showed morphogenetic responses, although shoot regeneration was mostly confined to primary calli, whereas rhizogenesis was more frequent in established calli. This indicates that not only hormonal combinations but also the differentiation stage of the tissue significantly affects regenerative outcomes.

Callus induction was efficiently achieved by Avenido and Hautea (1990) from cotyledon, epicotyl, and leaf explants of mungbean cultivars Pag-asa 2, Pag-asa 3, and accessions 363,354 and 250,162 when cultured on modified Murashige and Skoog (MS) medium containing 2.0 mg/L 2,4-D and 0.5 mg/L BA. Although this medium supported limited root formation (0–30%), transferring compact green calluses to MS medium supplemented with 2.0 mg/L BA, 0.1 mg/L NAA, and 1% sorbitol or 200 mg/L casein hydrolysate significantly enhanced rooting efficiency (57–95%). Similar study was performed by Patel et al. (1991) in MI7–21 and PIMS4 cultivars, where leaf explants from 8-day-old seedlings showed enhanced callus induction on MS medium supplemented with 6.7 μM benzyladenine (BA) and auxins, while kinetin was less effective which shows the versatility of mungbean explants and the importance of optimized hormonal treatments in achieving high-frequency regeneration for further genetic manipulation.

The medium containing BAP and kinetin was most favourable hormonal combinations for shoot regeneration from the proximal end of cotyledon (Gulati and Jaiwal 1990) and shoot tip explants (Gulati and Jaiwal 1992) of *Vigna radiata,* respectively. Addition of YE (100 mg/l) was found to be beneficial for shoot multiplication with Kn and NAA supplemented cultures of *Vigna radiata* (Mendoza and Futsuhara 1990). Gulati and Eapen (1992) used MS salts containing B5 vitamins to further study regeneration from shoot tip explants. Cytokinins like BA, kn, or zeatin (5×10^{-6} M) were used to promote multiple shoots in 100% of explants; BA produced the most shoots (up to 9 per explant). While adenine sulfate and NAA by themselves might induce full plantlets, auxin supplementation (NAA or IAA) could not improve shoot multiplication. The size of the shoot tip and the kind of cytokinin affected the effectiveness of regeneration. The 60% of the regenerated plants survived in the field after rooting on MS medium, either with or without auxins.

Plantlet regeneration through indirect organogenesis was obtained from leaf explants of *V. radiata* with BAP 1 mg/l (Mendoza et al. 1992a, b). Maximum shoot differentiation and elongation were observed even at low levels of TDZ while clusters of leafy structures appeared at the cotyledonary node region at higher levels in *Vigna radiata* (Sen and Guha-Mukherjee 1998). Tivarekar and Eapen (2001) reported high-frequency shoot regeneration protocol using immature cotyledons of *Vigna radiata* in BAP and IAA. Similar findings were also reported in matured mungbean explants cultured under in vitro condition (Chandra and Pal 1995)

The regenerative ability of hypocotyl, epicotyl, and cotyledon explants from three varieties of *Vigna radiata* was observed by Amutha et al. (2003); cotyledon cultures regenerated healthy shoots when compared to hypocotyl cultures. Thus, cotyledonary explants from mature seeds have been the most responsive explants in legumes for the induction of multiple shoots either directly from the surface of cotyledons or from nodal regions. However, these explants produced abnormal shoots when cultured in in vitro condition (Chandra and Pental 2003). Anatomical studies revealed the differences in the morphogenetic pattern formation from mature and immature cotyledons in different legumes. When mature cotyledons were cultured, either fresh shoots were induced from the proximal side of preexisting buds or shoot proliferation was observed around preexisting buds (Gulati and Jaiwal 1990; Chandra and Pal 1995).

An efficient regeneration protocol using immature cotyledons was described by Tivarekar and Eapen (2001), where the best shoot induction was achieved on MS medium supplemented with 2 mg/L BA and 0.5 mg/L IAA, while the addition of 0.5% activated charcoal completely inhibited shoot primordia formation; rooting was successfully induced with 0.1 mg/L IBA. In continuation, the light green cotyledon only produced normal shoots, while the light yellowish colored cotyledon did not produce normal shoots (Tivarekar and Eapen 2001). This study highlighted the importance of hormonal balance and explant maturity in achieving optimal regeneration. Similarly, effective direct shoot regeneration from cotyledonary node explants of both aromatic and non-aromatic types was documented by Betal and Sen Raychaudhuri (2001). Explants taken from 4-day-old seedlings with both cotyledons joined to the embryonic axis grew up to 5–6 shoots each on B5 basal medium supplemented with 3% sucrose and either BAP or 2-iP (5 μM).

High-frequency shoot regeneration from cotyledonary node and hypocotyl explants utilizing thidiazuron (TDZ) are also highlighted in other investigations. According to Amutha et al. (2006), abundant shoot regeneration (up to 104 shoots/explant from cotyledons) was the outcome of a 20-day initial exposure to TDZ (0.9 μM) followed by three subsequent subcultures with decreased TDZ (0.09 μM). Nevertheless, abnormalities brought on by TDZ, like fasciation and rosette leaf development, were observed. Mundhara and Rashid (2006) also observed that applying TDZ (1.0 μM) by flooding, an unusual but successful technique, increased shoot regeneration from cotyledonary nodes and promoted shoot production from the epicotyl.

Khatun et al. (2008) showed that in both the BINA mung 5 and BINA mung 7 types, cotyledon explants cultivated on MS media containing 1.0 mg/L BAP and

2.5 mg/L NAA showed the maximum callus induction (90%). Cotyledon-derived tissues showed a 62.5% success rate in callus regeneration on MS supplemented with 5.0 mg/L BAP and 0.05 mg/L NAA. The feasibility of shoot tip explants for direct organogenesis was further demonstrated by the 90% shoot differentiation shown by shoot tip explants grown directly in the same conditions in BINA mung 7. Vijayan et al. (2013) effectively developed an effective methodology for in vitro regeneration in cv. ML-267, despite the legume's resistance. The MS medium containing B5 vitamins and 5.0 mg/L BAP produced an average of 12.1 shoots per explant from 2-day-old cotyledonary node explants cultured under in vitro condition. Rooting was accomplished on MS medium containing 0.2 mg/L NAA with a 90% survival rate, and lower BAP concentrations promoted shoot elongation. High-frequency shoot regeneration from mungbean petiole explants on B5 medium containing BA was documented by Mahalakshmi et al. (2006a). Significantly more shoots (8–9 per explant) were produced by explants obtained from 10-day-old seedlings than by those derived from younger seedlings. Lamina promoted regeneration, and histological examination verified the shoots' adventitious origin. On B5 medium supplemented with 0.5 mg/L IBA, rooting was accomplished satisfactorily, and upon transfer to soil, a 90% survival rate was observed.

Mungbean regeneration is genotype dependent; hence every new cultivar must be optimized. The regeneration ability of the YMD-resistant cultivar 'Samrat' was assessed by Tripathi et al. (2022) utilizing explants of the embryonic axis (EA) and double cotyledonary node (DCN). EA explants yielded more shoots (6.45) on MS media with 1 mg/L BAP than DCN (5.24) (Table 1). Following three subcultures on BAP-supplemented media and two on basic MS medium, the best results were obtained. For EA, rooting was accomplished without growth regulators, and for DCN, on 1.0 mg/L IAA. 72–80% of plantlets survived in a greenhouse.

Surprisingly, higher levels of micronutrients, carbohydrates, or nitrogen than the typical B5 composition did not promote shoot multiplication, indicating that hormonal balance rather than nutrient enrichment was the primary factor. Similar study was done by Rathore et al. (2023), shoots regenerated were higher with embryonic axis (4.02) than with double cotyledonary node (3.1).

These studies highlight the importance of culture conditions, cytokinin concentration, genotype, and explant type in affecting the effectiveness of mungbean in vitro shoot regeneration. Explants of shoot tips and cotyledonary nodes, especially from early seedlings, respond very well to the right cytokinin concentrations in culture. The techniques that have been presented provide important information for biotechnological uses including genome editing and genetic transformation in mungbean enhancement programs.

2.2 In Vitro Regeneration Through Somatic Embryogenesis

Somatic embryos morphologically resemble zygotic embryos except the fact that they develop embedded in maternal tissues. Somatic embryogenesis can be achieved through optimization of various culture conditions and medium combinations.

Table 1 In vitro plant regeneration in mungbean

Sl. No	Explants	Medium composition	Regeneration methods	Reference
1.	Cotyledon, epicotyl, and leaf explants	MS + BA (1.0 mg/L)	Indirect organogenesis	Avenido and Hautea (1990)
2.	Proximal ends of cotyledons	MS + BAP (1×10^{-1} M)	Direct organogenesis	Gulati and Jaiwal (1990)
3.	Shoot tips	MS + BA (5×10^{-6} M)	Direct organogenesis	Gulati and Jaiwal (1992)
4.	Cotyledonary node	B_5 basal medium (BBM) + BAP or 2-iP, respectively (5×10^{-7}M)	Direct organogenesis	Gulati and Jaiwal (1994)
5.	Cotyledons	MS + BA	Direct organogenesis	Chandra and Pal (1995)
6.	Immature cotyledon	MS liquid medium containing 2,4-D (1.5 mg/L) and L-proline (50 mg/L)	Somatic embryogenesis	Girija et al. (2000)
7.	Immature cotyledon	MS + BA (2 mg/L) + IAA (0.5 mg/L)	Direct organogenesis	Tivarekar and Eapen (2001)
8.	Hypocotyl and primary leaves	MS + BAP (5×10^{-6} M+ 2.5×10^{-6} M each of 2,4-D and NAA	Direct organogenesis	Jaiwal et al. (2001)
9.	Hypocotyl	MS + BA (6.66 μm) + TDZ (2.5 μm) +10% coconut water	Indirect organogenesis	Amutha et al. (2003)
10.	Mature cotyledons, hypocotyl, nodal segment, and leaf	MS + 2,4-D (1.809 mM) + BA (3.555 Mm)	Somatic embryogenesis	Devi et al. (2004)
11.	Cotyledonary node and hypocotyl	MS + TDZ (0.9 μM)	Direct and indirect organogenesis	Amutha et al. (2006)
12.	Cotyledonary node	MS + TDZ (1.0 μM)	Direct organogenesis	Mundhara and Rashid (2006)
13.	Cotyledon	BAP (2.22–8.88 Mm) or Kin (2.32–9.38 μM)	Somatic embryogenesis	Kaviraj et al. (2006)
14.	Cotyledon and shoot tip	MS + BAP (5 mg/L) + NAA (0.05 mg/L)	Direct organogenesis	Khatun et al. (2008)
15.	Cotyledon attached with embryonal axis	MS + BAP (1.0 mg/L)	Direct organogenesis	Islam and Islam (2010)
16.	Primary leaf	MS+ BA (0.5 mg/L) and KNO_3 (500 mg/L)	Somatic embryogenesis	Sivakumar et al. (2010)
17.	Cotyledonary node	MS salts, B5 vitamins +BAP (5.0 mg/L)	Direct organogenesis	Vijayan et al. (2013)

(continued)

Table 1 (continued)

Sl. No	Explants	Medium composition	Regeneration methods	Reference
18.	Cotyledonary node		Direct organogenesis	Sahoo et al. (2016)
19.	Cotyledonary node, shoot tip, and cotyledon-attached decapitated embryo	MS + BAP (10.0 mM) + NAA (0.5 μM)	Direct organogenesis	Begum (2017)
20.	Cotyledonary node	B5 + BAP (1.0 μm)	Direct organogenesis	Baloda et al. (2017)
21.	Cotyledonary node	MSB5 + BAP (1.0 μM)	Direct organogenesis	Kumar et al. (2017)
22.	Cotyledon-attached decapitated embryo	MS + BAP (4.0 μM) + NAA (0.5 μM)	Direct organogenesis	Bhajan et al. (2019)
23.	Hypocotyl	BAP (1 μM) + GA_3 (12 μM)	Somatic embryogenesis	Cholke Pravin et al. (2019)
24.	Cotyledon	MS+ BAP (1 μM) + GA_3(12 M)	Somatic embryogenesis	Gnanaraj et al. (2022)
25.	Double Cotyledonary Node (DCN) and Embryonic Axis (EA)	MS + BAP (1.0 mg/L)	Direct organogenesis	Tripathi et al. (2022)
26.	Cotyledonary node	MS + BAP (5 mg/L)	Direct organogenesis	Ramachandra et al. (2022)
27.	Double Cotyledonary Node (DCN) and Embryonic Axis (EA)	MS + BAP (1.0 mg/L)	Direct organogenesis	Rathore et al. (2023)

In many legumes, particularly mungbean, direct somatic embryogenesis was observed from immature explants (Girija et al. 2000; Sivakumar et al. 2010; Cholke Pravin et al. 2019), whereas highly differentiated explants usually undergo a callus phase to acquire embryogenic potential (Prakash et al. 1994; Devi et al. 2004; Kaviraj et al. 2006; Sivakumar et al. 2010, Sindhujaa et al. 2018) (Table 1).

Somatic embryogenesis was induced from immature cotyledon and embryonic axis, mature stages could not be achieved, and most of them were abnormal (George and Eapen 1994). Somatic embryogenesis was successfully achieved using various explants, culture media, and growth regulators. Immature cotyledon-derived calli were used to start somatic embryogenesis on MS medium that was supplemented with NAA (5 mg/L). Somatic embryos underwent differentiation after being transferred to MS liquid medium with 2,4-D (1.5 mg/L) and L-proline (50 mg/L). MS media supplemented with BAP and ABA (1 mg/L each) were used to increase the maturation and germination of torpedo-stage embryos, leading to a 50% conversion rate to plantlets. Half-strength MS solid medium was used to continue develop these germinated embryos (Girija et al. 2000). Devi et al. (2004) also used different explants like mature cotyledons, hypocotyl, nodal segment, and leaf explants for somatic embryo induction in MS medium containing 1.809 μM of 2,4-D and 3.555 μM BA. They successfully developed plants with survival rate of 92%.

Kaviraj et al. (2006) produced normal shape of cotyledonary embryos from mature cotyledon explants of var. TAP-7. The mature embryos were normally germinated in MS medium containing 1.88 μM ABA and 6.66 μM BAP with regeneration frequency of 56.6%. Similarly, Sivakumar et al. (2010) successfully developed the plantlets from mature somatic embryos cultured in MS medium containing 0.5 mg/L BA and 500 mg/L KNO_3. In contrast, Cholke Pravin et al. (2019) produced somatic embryos from hypocotyl explants cultured in MS medium containing 50 mg/L proline and 2 mg/L 2, 4 D. Gnanaraj et al. (2022) established a somatic embryogenesis protocol using PEG as an osmotic agent. Embryogenic calli induced with 5 μM picloram developed into torpedo and cotyledonary embryos in ABA- and proline-supplemented media. Plantlets were regenerated with 1 μM BAP + 12 μM GA_3 and showed 99% survival after transfer to field conditions (Table 1).

3 Genetic Transformation in Mungbean

There are only a few reports available on genetic transformation studies of mungbean. Peychoknagul et al. (1996) made transformation using cotyledon explants of *Vigna radiata* through *Agrobacterium*-mediated transformation and confirmed by PCR and Southern blot hybridization. Jaiwal et al. (2001) for the first time confirmed the stable expression of transgene in transformed shoots from cotyledonary node explants of *Vigna radiata*. However, they failed to regenerate shoots from transformed calli derived from hypocotyl and primary leaves.

Several studies have successfully established *Agrobacterium*-mediated genetic transformation protocols in mungbean, employing a variety of explants and transformation parameters. Tazeen and Mirza (2004) reported the highest transformation efficiency (80%) using primary leaves by using *Agrobacterium* strain C58C1 carrying the p35SGUSINT vector. Similarly, Jaiwal et al. (2001) used several *Agrobacterium* strains (LBA4404, EHA105, and C58C1) to show that cotyledonary node explants not only produced green shoots with high GUS activity, but also regenerated fertile, transgene-integrated plants, as confirmed by Southern blotting. This comparison demonstrates the effect of both bacterial strain and explant type on transformation efficiency. Begum (2017) and Bhajan et al. (2019) found that the cotyledon-attached decapitated embryo and cotyledonary node explants were highly responsive to shoot induction and transformation. The stable integration of GUS and nptII genes was confirmed by PCR (Table 2).

Building on these findings, Ramachandra et al. (2022) studied the transformation of diverse mungbean varieties including KKM-3 and China mung using *Agrobacterium* LBA4404 harboring pCAMBIA1305.2 with MYMV Rep gene, where shoots were regenerated on MS medium containing 5 mg/L BAP, elongated with zeatin and BAP, and rooted on ½ MS with NAA, again confirmed through GUS assay and PCR. Their findings supported the importance of genotype-specific responses and post-transformation regeneration steps in successful plant recovery. Suraninpong et al. (2011) used both *A. tumefaciens* and *A. rhizogenes* strains and achieved up to 31.25% transformation efficiency from BAP-treated cotyledon

Table 2 Genetic transformation studies in mungbean

Sl. No	Explants	Gene	Trait	Method	Reference
1.	Cotyledons	*GUS*	–	*Agrobacterium rhizogenes*	Suraninpong et al. (2011)
2.	Cotyledonary leaf and cotyledon attached with embryonal axis	*GUS* and *nptII*	–	*Agrobacterium*	Islam and Islam (2010)
3.	Hypocotyl and primary leaves	*GUS* and *nptII*	–	*Agrobacterium*	Jaiwal et al. (2001)
4.	Cotyledon-attached decapitated embryo, cotyledonary node, and decapitated embryo	*GUS* and *nptII*	–	*Agrobacterium*	Begum (2017)
5.	Cotyledon-attached decapitated embryo	*GUS* and *nptII*	–	*Agrobacterium*	Bhajan et al. (2019)
6.	Primary leaf	*p35SGUSINT GUS* and *nptII*	–	*Agrobacterium*	Tazeen and Mirza (2004)
7.	Primary leaf explant	*GUS* and *hpt1*	–	*Agrobacterium*	Mahalakshmi et al. (2006b)
8.	Cotyledonary node	*AtNHX1*	Salinity	*Agrobacterium*	Sahoo et al. (2016)
9.	Cotyledonary node	*codA a*	Salinity	*Agrobacterium*	Baloda et al. (2017)
10.	Cotyledonary node	*AtNHX1* and *bar*	Salinity, oxidative stress, and herbicide	*Agrobacterium*	Kumar et al. (2017)
11.	Cotyledonary nodes	MYMV Replicase (Rep)	MYMV	*Agrobacterium*	Ramachandra et al. (2022)
12.	Embryonic axis	*AtDREB1a*	–	*Agrobacterium* and Biolistic method	Tripathi et al. (2024)

explants. Their research identified factors such as seedling age, genotype, and co-cultivation parameters as critical to reproducibility. Taken together, these studies show that mungbean transformation success is determined by a synergistic combination of explant type, bacterial strain, vector system, selection pressure, and genotype responsiveness, with reliable molecular confirmation methods like GUS staining, PCR, and Southern blotting (Table 2).

Current genetically modified research in mungbean has focused on increasing resistance to abiotic stress by the designed insertion of particular candidate genes. Incorporating the traditional methods of transformation, researchers have probed and validated the gene role in reaction to salinity (Baloda et al. 2017; Kumar et al. 2017; Sahoo et al. 2016) and cold stress (Rout et al. 2020). Specifically, Kumar et al.

(2017) and Sahoo et al. (2016) overexpressed *Arabidopsis* genes, indicating how genetic enhancements can be developed at accelerated rates when using insights gained in model organisms to target genetic improvements in mungbean. Genome-editing instruments have been utilized in mungbean in the recent years, especially in focus on biotic stress (Talakayala et al. 2022). One of the most prominent benefits of genome editing, in particular through gene knockout approaches that fall under SDN1 options, lies in their categorization as non-genetically modified organisms (GMOs) in various regulatory environments (Zhang et al. 2021).

Although editing the genome seems like a clear benefit to transfer better abiotic stress levels, knock-out approach does not always provide beneficial physiological outcomes. A complementary approach thus involves altering gene regulatory elements, which can bring about a broader range of phenotypic variations (Crisp et al. 2022). However, genome editing continues to be a useful tool in functional genomics and in interrogating well-defined genes in other plant species. High-throughput editing, implemented in commercial mungbean varieties, can have the same effect; deploying stress-resistant varieties to farmers would occur faster. Furthermore, a study demonstrated the utility of both *Agrobacterium*-mediated and biolistic transformation methods in the cultivar Samrat. The biolistic transformation of embryonic axis explants using PEG-induced drought stress yielded a transformation efficiency of 14.22%, while the *Agrobacterium* method yielded 1.19%. Transgene inheritance and expression were confirmed through PCR and Southern blot analysis across T_1 to T_3 generations. These advancements highlight the growing potential of mungbean as a transgenic platform for abiotic stress resilience, paving the way for climate-resilient crop development through targeted genetic manipulation.

4 Conclusion

Over the last few decades, in vitro regeneration and genetic transformation of the crop *V. radiata* established the crop as a potential model to apply biotechnological interventions to advance the crop as well as legume as a whole. Broad explant competence has been demonstrated, as cotyledonary nodes, shoot tips, hypocotyls, and embryonic axes have exhibited regenerative potential when subjected to optimized hormonal conditions. The most effective treatments involved cytokinins, particularly BAP and TDZ. The somatic embryogenesis pathway is a valid alternative regeneration to the conventional organogenesis process, particularly to recalcitrant genotypes. Although there are progress, regeneration efficiencies and transformation success are severely genotype-limiting often preventing reproducibility and scale-up of protocols to other cultivars. Recent integration of tools such as marker-assisted selection allows more precise optimization, while transgene confirmation using PCR, GUS assays, and Southern blotting provides a more accurate validation approach, thereby increasing confidence in transformation systems. Technologies emerging in genome editing are other potential approaches that can lead to trait enhancement and particularly in abiotic stress tolerance. Nevertheless, such limitations as low frequencies of transformation and somaclonal variation should still have to be overcome. The future directions of research ought to be based on establishing genotype-independent processes, identifying optimal liquid culture systems based on bioreactor, and using precision editing techniques. All these developments have he capacity to accelerate the genetic improvement of mungbean and enhance its contribution to food and nutritional security in the world.

References

Amutha S, Ganapathi A, Muruganantham M (2003) In vitro organogenesis and plant formation in *Vigna radiata* (L.) Wilczek. Plant Cell Tissue Organ Cult 72:203–207. https://doi.org/10.1023/A:1022266110750

Amutha S, Muruganantham M, Ganapathi A (2006) Thidiazuron-induced high-frequency axillary and adventitious shoot regeneration in *Vigna radiata* (L.) Wilczek. In Vitro Cell Dev Biol Plant 42:26–30. https://doi.org/10.1079/IVP2005721

Avenido RA, Hautea DM (1990) In vitro organogenesis and flowering in Mungbean (*Vigna radiata* L. Wilczek). Philipp J Crop Sci 15(3):169–173

Baloda A, Madanpotra S, Jaiwal PK (2017) Transformation of Mungbean plants for salt and drought tolerance by introducing a gene for an osmoprotectant glycine betaine. J Plant Stress Physiol 3:5–11

Baza M, Shanka D, Bibiso M (2022) Agronomic and economic performance of Mungbean (*Vigna radiata* L.) varieties in response to rates of blended NPS fertilizer in KindoKoysha district, Southern Ethiopia. Open Life Sci 17(1):1053–1063. https://doi.org/10.1515/biol-2022-0461

Begum S (2017) Development of in vitro plant regeneration system and agrobacterium–mediated genetic transformation in Mungbean (*Vigna radiata* (L.) wilczeck) (Doctoral dissertation, University of Dhaka)

Betal S, Sen Raychaudhuri S (2001) Shoot tip multiplication in aromatic and non-aromatic varieties of *Vigna radiata* (L.) Wilczek. Plant Cell Tissue Organ Cult (PCTOC) 11:187–194

Bhajan SK, Hasan MM, Rahman MA, Sarker RH, Islam MN (2024) An efficient approach of in vitro plant regeneration and propagation of Mungbean [*Vigna radiata* (L.) Wilczek]. Plant Trends 2(3):46. https://doi.org/10.5455/pt.2024.05

Bhajan SK, Begum S, Islam MN, Hoque MI, Sarker RH (2019) In vitro regeneration and Agrobacterium-mediated genetic transformation of local varieties of Mungbean (*Vigna radiata* (L). Wilczek). Plant Tissue Cult Biotechnol 29(1):81–97. https://doi.org/10.3329/ptcb.v29i1.41981

Bhatia S, Sharma K, Dahiya R, Bera T (2015) Modern applications of plant biotechnology in pharmaceutical sciences. Academic

Chandra A, Pental D (2003) Regeneration and genetic transformation of grain legumes: an overview. Curr Sci 84(3):381–387

Chandra M, Pal A (1995) Differential response of the two cotyledons of *Vigna radiata* in vitro. Plant Cell Rep 15:248–253. https://doi.org/10.1007/BF00193729

Cholke Pravin B, Bhor Abhishek K, Shete Akshay M (2019) Somatic embryogenesis in mung beans (*Vigna radiata* L.)

Crisp PA, Bhatnagar-Mathur P, Hundleby P, Godwin ID, Waterhouse PM, Hickey LT (2022) Beyond the gene: epigenetic and cis-regulatory targets offer new breeding potential for the future. Curr Opin Biotechnol 73:88–94. https://doi.org/10.1016/j.copbio.2021.07.008

Dahiya PK, Linnemann AR, Van Boekel MA, Khetarpaul N, Grewal RB, Nout MJ (2015) Mungbean: technological and nutritional potential. Crit Rev Food Sci Nutr 55(5):670–688. https://doi.org/10.1080/10408398.2012.671202

Devi P, Radha P, Sitamahalakshmi L, Syamala D, Kumar SM (2004) Plant regeneration via somatic embryogenesis in Mungbean [*Vigna radiata* (L.) Wilczek]. Sci Hortic 99(1):1–8. https://doi.org/10.1016/S0304-4238(03)00079-7

George L, Eapen S (1994) Organogenesis and embryogenesis from diverse expiants in pigeonpea (*Cajanus cajan* L. Millsp.). Plant Cell Rep 13:417–420

Girija S, Ganapathi A, Ananthakrishnan G (2000) Somatic embryogenesis in *Vigna radiata* (L.) Wilczek. Indian J Exp Biol 38(12):1241–1244

Gnanaraj M, Sneka C, Sisubalan N, Baburajan R, Manikandan R, Muneeswaran T (2022) Polyethylene glycol induced somatic embryogenesis and plant regeneration from cotyledons of *Vigna radiata* (L.) Wilczek. S Afr J Bot 150:721–730

Gulati A, Jaiwal PK (1990) Culture conditions effecting plant regeneration from cotyledons of *Vigna radiata* (L.) Wilczek. Plant Cell Tissue Organ Cult 23:1–7. https://doi.org/10.1007/BF00116082

Gulati A, Jaiwal PK (1992) In vitro induction of multiple shoots and plant regeneration from shoot tips of Mungbean (*Vigna radiata* (L.) Wilczek). Plant Cell Tissue Organ Cult 29:199–205. https://doi.org/10.1007/BF00034353

Gulati A, Jaiwal PK (1994) Plant regeneration from cotyledonary node explants of Mungbean (*Vigna radiata* (L.) Wilczek). Plant Cell Rep 13:523–527. https://doi.org/10.1007/BF00232949

Hesami M, Pepe M, Spitzer-Rimon B, Eskandari M, Jones AM (2024) Epigenetic factors related to recalcitrance in plant biotechnology. Genome 68:1–11. https://doi.org/10.1139/gen-2024-0098

Islam MN, Islam KT (2010) Agrobacterium-mediated genetic transformation of Mungbean (*Vigna radiata* (L.) Wilczek). Plant Tissue Cult Biotechnol 20(2):233–236

Jaiwal PK, Kumari R, Ignacimuthu S, Potrykus I, Sautter C (2001) *Agrobacterium* tumefaciens-mediated genetic transformation of Mungbean (*Vigna radiata* L. Wilczek) – a recalcitrant grain legume. Plant Sci 161(2):239–247. https://doi.org/10.1016/S0168-9452(01)00352-1

Kaviraj CP, Kiran G, Venugopal RB, Kavi Kishor PB, Rao S (2006) Somatic embryogenesis and plant regeneration from cotyledonary explants of green gram [*Vigna* radiata (L.) Wilezek.] – a recalcitrant grain legume. In Vitro Cell Dev Biol Plant 42(2):134–138. https://doi.org/10.1079/IVP2005739

Khatun MK, Haque MS, Islam S, Nasiruddin KM (2008) In vitro regeneration of Mungbean (*Vigna radiata* L.) from different explants. Progress Agric 19(2):13–19

Kumar S, Kalita A, Srivastava R, Sahoo L (2017) Co-expression of Arabidopsis NHX1 and bar improves the tolerance to salinity, oxidative stress, and herbicide in transgenic Mungbean. Front Plant Sci 8:1896. https://doi.org/10.3389/fpls.2017.01896

Loberant B, Altman A (2010) Micropropagation of plants. In: Encyclopedia of industrial biotechnology: bioprocess, bioseparation, and cell technology. Wiley, New York, p 34993515

Mahalakshmi LS, Leela T, Kumar BK, Naresh B, Devi P (2006a) In vitro plant regeneration from the petioles of primary leaves of Mungbean *Vigna radiata* L. Plant Biotechnol 23(4):409–411. https://doi.org/10.5511/plantbiotechnology.23.409

Mahalakshmi LS, Leela T, Kumar SM, Kumar BK, Naresh B, Devi P (2006b) Enhanced genetic transformation efficiency of Mungbean by use of primary leaf explants. Curr Sci 91(1):93–99

Mathews H (1987) Morphogenetic responses from in vitro cultured seedling explants of Mungbean (*Vigna radiata* L. Wilczek). Plant Cell Tissue Organ Cult 11:233–240. https://doi.org/10.1007/BF00040429

Mathews VH, Rao PS (1984) In vitro production of multiple seedlings from single seeds of Mungbean (*Vigna radiata* L. Wilczek). Z Pflanzenphysiol 113(4):325–329. https://doi.org/10.1016/S0044-328X(84)80038-0

Mendoza AB, Futsuhara Y (1990) Varietal differences on plant regeneration by tissue culture in mungbean Vigna radiata (L.) Wilczek. Japan J Breed 40: 457–467

Mendoza AB, Hattori K, Futsuhara Y (1992a) Shoot regeneration from the callus of immature primary leaves in Mungbean (*Vigna radiata* (L.) Wilczek). Jpn J Breed 42:145–149

Mendoza AB, Kazumi H, Futsuhara Y (1992b) Shoot regeneration from the callus of immature primary leaves in mungbean. J Breed 42(1):145–149

Mundhara R, Rashid A (2006) Recalcitrant grain legume *Vigna radiata*, Mungbean, made to regenerate on change of hormonal and cultural conditions. Plant Cell Tissue Organ Cult 85:265–270. https://doi.org/10.1007/s11240-005-9061-3

Nivya VM, Shah JM (2023) Recalcitrance to transformation, a hindrance for genome editing of legumes. Front Genome Edit 5:1247815. https://doi.org/10.3389/fgeed.2023.1247815

Pataczek L, Zahir ZA, Ahmad M, Rani S, Nair R, Schafleitner R, Cadisch G, Hilger T (2018) Beans with benefits—the role of Mungbean (*Vigna radiata*) in a changing environment. Am J Plant Sci 9(07):1577. https://doi.org/10.4236/ajps.2018.97115

Patel MB, Bharadwaj R, Joshi A (1991) Organogenesis in *Vigna radiata* (L.) Wilczek. Indian J Exp Biol 29:619–622

Peychoknagul S, Phonjun C, Pongtongkam P, Ngernsiri L (1996) Gene transformation of mungbean. Kasetsart J (Nat Sci) 30:303–311

Prakash NS, Pental D, Bhalla-Sarin N (1994) Regeneration of pigeonpea (Cajanus cajan L.) from cotyledonary nodes via multiple shoot formation. Plant Cell Rep 13:623–627

Ramachandra A, Roshni M, Anitha P, Bhat SG, Raviteja C (2022) Isolation, cloning and agrobacterium mediated transformation of MYMV replicase (Rep) gene in *Vigna radiata* (L.) Wilczek for MYMV resistance. Asian J Microbiol Biotechnol Environ Sci 24(1):102–107. https://doi.org/10.53550/AJMBES.2022.v24io1.019

Rathore M, Tripathi A, Kushwah NS, Singh NP (2023) Multiple shoot regeneration from detached embryonic Axis in Greengram (*Vigna radiata*) cv. SML 668. Legume Res-Int J 46(3):324–330. https://doi.org/10.18805/LR-4693

Rout GR, Bansal A, Swain D, Jadhao KR, Shelke RG, Panda SK (2020) Overexpression of ICE1 gene in Mungbean (*Vigna radiata* L.) for cold tolerance. Plant Cell Tissue Organ Cult 143:593–608. https://doi.org/10.1007/s11240-020-01944-w

Sahoo DP, Kumar S, Mishra S, Kobayashi Y, Panda SK, Sahoo L (2016) Enhanced salinity tolerance in transgenic Mungbean overexpressing Arabidopsis antiporter (NHX1) gene. Mol Breed 36:1–15. https://doi.org/10.1007/s11032-016-0564-x

Sen J, Guha-Mukherjee S (1998) In vitro induction of multiple shoots and plant regeneration in Vigna. In Vitro Cell Dev Biol Plant 34(4):276–280

Sindhujaa V, Gnanaraj M, Viji M, Karuppanapandian T, Manoharan K (2018) Induction of high frequency somatic embryogenesis and analysis of developmental stagewise expression of SERK1 gene during somatic embryogenesis in cultures of *Vigna* radiata (L.) R. Wilczek. Indian J Exp Biol 56(3):180–193

Sivakumar P, Gnanam R, Ramakrishnan K, Manickam A (2010) Somatic embryogenesis and regeneration of *Vigna radiata*. Biol Plant 54:245–251. https://doi.org/10.1007/s10535-010-0043-x

Suraninpong P, Chanprame S, Cho HJ, Widholm JM, Waranyuwat A (2011) *Agrobacterium*-mediated transformation of Mungbean [*Vigna radiata* (L.) Wilczek]. Walailak J Sci Technol 1(2):38–48. https://wjst.wu.ac.th/index.php/wjst/article/view/184

Swamy KRM (2023) Origin, domestication, taxonomy, botanical description, genetics and cytogenetics, genetc diversity and breeding of Mungbean (*Vigna radiata* (L.) R. Wilczek). Int J Curr Res 15(7):25351–25375

Talakayala A, Mekala GK, Malireddy RK, Ankanagari S, Garladinne M (2022) Manipulating resistance to Mungbean yellow mosaic virus in greengram (*Vigna radiata* L): through CRISPR/Cas9 mediated editing of the viral genome. Front Sustain Food Syst 6:911574. https://doi.org/10.3389/fsufs.2022.911574

Tazeen S, Mirza BU (2004) Factors affecting Agrobacterium tumefaciens mediated genetic transformation of *Vigna radiata* (L.) Wilczek. Pak J Bot 36(4):887–896

Tivarekar S, Eapen S (2001) High frequency plant regeneration from immature cotyledons of Mungbean. Plant Cell Tissue Organ Cult 66:227–230. https://doi.org/10.1023/A:1010644812268

Tripathi A, Kushwah NS, Debnath SC, Shukla S, Rathore M (2022) Investigating the in vitro regeneration potential of Mungbean cultivar Samrat. J Food Legumes 35(3):198–206. https://doi.org/10.59797/journaloffoodlegumes.v35i3.331

Tripathi A, Rathore M, Shukla S, Das A, Debnath SC (2024) Agrobacterium and biolistic mediated genetic transformation of Mungbean cultivar Samrat using embryogenic explant. Plant Cell Tissue Organ Cult 157(3):72

Vijayan S, Beena MR, Kirti PB (2006) Simple and effective regeneration of Mungbean (*Vigna radiata* (L) Wilczek) using cotyledonary node explants. J Plant Biochem Biotechnol 15:131–134. https://doi.org/10.1007/BF03321916

Yadav SK, Krishna MG, Maheswari M, Vanaia M, Venkateswarlu B (2010) High frequency induction of multiple shoots and plant regeneration from cotyledonary nodal explant of Mungbean [*Vigna radiata* (L) Wilczek]. J Plant Biochem Biotechnol 19(2):267–270. https://doi.org/10.1007/BF03263354

Yousif DP (2020) Improving of Mungbean production using nuclear techniques and package of technology with mutation induction and tissue culture. Concepts Dairy Vet Sci 3(3):311–312

Zhang Y, Restall J, Crisp P, Godwin I, Liu G (2021) Current status and prospects of plant genome editing in Australia. In Vitro Cell Dev Biol Plant 57(4):574–583. https://doi.org/10.1007/s11627-021-10188-y

Gene Editing in Mungbean: Opportunities and Challenges

Jothi Kanmani Bharathi, Sanyasi Selvamani, Sivakumar Paramasivam, and Shen Wang

Abstract

Mungbean (*Vigna radiata*), a nutritionally rich legume extensively cultivated in Asia, has tremendous potential for sustainable agriculture due to its short growth cycle, nitrogen-fixing capabilities, and adaptability to varied temperatures. Traditional breeding methods have improved mungbean varieties but are hampered by a narrow genetic basis, complicated multi-gene characteristics, and susceptibility to biotic and abiotic stressors, hampering efficient trait development. This chapter discusses the transformative impact of modern genomic techniques, particularly whole genome sequencing (WGS) and CRISPR/Cas9 gene editing, in overcoming these hurdles. WGS has explained mungbean's genomic architecture, finding essential genes for features such as bruchid resistance, powdery mildew resistance, seed size, and plant architecture, while transcriptomics has revealed molecular responses to stressors. CRISPR/Cas9, successfully employed in model legumes and food crops, allows precise phenotypic modification, building on knowledge translatable to mungbean. However, problems such as transformation recalcitrance and off-target consequences limit progress. Low transformation efficiencies (e.g., 4.2% in cotyledonary nodes) and regeneration

J. Kanmani Bharathi (✉)
Faculty of Agriculture, Annamalai University, Annamalainagar, Tamil Nadu, India

S. Selvamani
Tamil Nadu Agricultural University, Coimbatore, Tamil Nadu, India

S. Paramasivam
Dr.M.S.Swaminathan Agricultural College and Research Institute, Tamil Nadu Agricultural University, Thanjavur, Tamil Nadu, India

S. Wang
Institute of Cereal and Oil Crops, Hebei Academy of Agricultural and Forestry Sciences/Hebei Laboratory of Crop Genetic and Breeding, Shijiazhuang, Hebei, China

S. Paramasivam et al. (eds.), *Mungbean*,
https://doi.org/10.1007/978-981-95-7637-1_10

challenges need creative techniques, including tissue-specific promoters, high-fidelity Cas9 variants, and alternative delivery mechanisms such as ribonucleoproteins (RNPs) and viral vectors. Optimizing guide RNA design, utilizing base and prime editing, and harnessing growth-promoting factors further boost editing precision and efficiency. Virus-mediated and nanotechnology-based techniques give transgene-free options, boosting heredity. By merging these breakthroughs with traditional breeding, mungbean improvement can be hastened, creating durable, high-yielding variants to address global food security needs. This analysis underlines the necessity for continuing research to fully utilize these technologies for mungbean's sustainable genetic improvement.

Keywords

Bruchid · Gene editing · Powdery mildew · transformation

1 Introduction

Mungbean (*Vigna radiata*), commonly referred to as green gram, is a significant legume crop belonging to the Fabaceae family. It is primarily cultivated in Asia, particularly in countries such as India, Pakistan, Thailand, and China, and has gained traction in regions like Australia and the United States due to its nutritional benefits and adaptability to various climates (Aneja et al. 2013; Pratap et al. 2021). Mungbean is characterized by its short growth duration, typically around 60 days, making it an attractive option for farmers seeking quick returns on their investments (Aziz-ur-Rehman et al. 2019).Nutritionally, mungbean is an excellent source of protein, containing approximately 20–30% protein and 50–60% carbohydrates, along with essential vitamins and minerals such as folate and iron (Yang et al. 2013). This high nutritional value positions mungbean as a vital dietary component, particularly in regions where protein sources are limited (Yi-Shen et al. 2018). The seeds can be consumed in various forms, including whole, sprouted, or processed into flour, noodles, and protein concentrates, which further enhances its versatility in culinary applications (Dahiya et al. 2015). The cultivation of mungbean also plays a crucial role in sustainable agriculture due to its ability to fix atmospheric nitrogen, thereby improving soil fertility and reducing the need for chemical fertilizers (Nair et al. 2014; Pataczek et al. 2018). This characteristic is particularly beneficial in intercropping systems, where mungbean can be rotated with staple crops like rice and wheat, contributing to a more sustainable agricultural practice (Diatta et al. 2024). Furthermore, mungbean's drought tolerance and low input requirements make it suitable for cultivation in marginal lands, thus supporting food security in developing regions (Liang et al. 2020). Genetic diversity within mungbean is significant, with various landraces and improved varieties developed to enhance yield and disease resistance (Tripathi et al. 2020). Recent advancements in genomic research have provided insights into the genetic makeup of mungbean, facilitating breeding programs aimed at improving traits such as disease resistance and stress tolerance

(Somta et al. 2022). For instance, studies have identified genetic markers associated with resistance to pests and diseases, which are critical for developing resilient mungbean varieties (Singh et al. 2022a, b, c). Despite its advantages, mungbean cultivation faces challenges, including susceptibility to various pests and diseases, such as the mungbean yellow mosaic virus and bruchid beetles, which can significantly impact yields. Integrated pest management strategies and the development of resistant varieties are essential for mitigating these threats and ensuring sustainable production (Nair et al. 2019). Moreover, the identification of quantitative trait loci (QTLs) associated with important agronomic traits, such as seed size and disease resistance, has been reported, providing valuable insights for marker-assisted selection (Sompong et al. 2012). In addition to molecular markers, genomic resources have been developed to support breeding efforts. The near-complete genome sequence of mungbean has been established, which aids in understanding the genetic basis of various traits and facilitates the identification of candidate genes for specific improvements (Ha et al. 2021). Furthermore, the exploration of genetic variability within the mungbean gene pool is essential for breeding programs. The conservation and characterization of diverse landraces and wild relatives of mungbean provide a reservoir of genetic traits that can be integrated into cultivated varieties (Manjunatha et al. 2023). Moreover, the application of advanced breeding techniques, such as genomic selection and CRISPR/Cas9 gene editing, holds promise for accelerating the genetic improvement of mungbean. These technologies enable precise modifications of specific traits, leading to the development of high-yielding and stress-resistant varieties (Singh et al. 2017). The integration of these modern techniques with traditional breeding methods can significantly enhance the efficiency of mungbean breeding programs. Continued research and investment in genetic improvement efforts will be essential to ensure the sustainability and profitability of mungbean cultivation, ultimately contributing to food security and nutritional enhancement in regions where this crop is a staple.

2 Mungbean Genetic Background and Improvement Strategies

Genetic diversity and breeding efforts in mungbean are critical for enhancing the crop's resilience, yield, and nutritional quality. Mungbean is a self-pollinated legume, which inherently limits its genetic diversity due to the restricted gene flow between varieties. This narrow genetic base poses challenges for breeding programs aimed at improving traits such as disease resistance, drought tolerance, and yield potential (Gwag et al. 2010). Consequently, there is a pressing need to broaden the genetic base of cultivated mungbeans through various breeding strategies. One effective approach to enhancing genetic diversity is the introgression of genes from wild relatives and other *Vigna* species. Studies have shown that hybridization between *Vigna radiata* and its relatives, such as *Vigna mungo* and *Vigna umbellata*, can successfully introduce desirable traits, including resistance to pests and diseases (Pratap et al. 2021). Molecular markers, particularly simple sequence repeats (SSR), have been instrumental in assessing genetic diversity among mungbean accessions. The application of SSR markers has enabled researchers to detect genetic variability and identify improved varieties that are genetically similar to landraces, thereby supporting breeding programs and conservation strategies (Reflinur et al. 2016). Moreover, genetic linkage maps constructed using both cultivated and wild mungbean accessions have provided valuable insights into the inheritance of important traits, facilitating marker-assisted selection (Isemura et al. 2012). The World Vegetable Center (AVRDC) has played a pivotal role in mungbean breeding efforts, releasing over 110 improved cultivars that have significantly increased global production (Nair et al. 2014). These cultivars are characterized by short growth durations and enhanced resistance to biotic stresses, which are essential for adapting to diverse cropping systems and climatic conditions (Nair et al. 2019). Furthermore, the integration of biotechnological approaches, such as *Agrobacterium*-mediated transformation, has opened new avenues for introducing specific traits into mungbean varieties, thereby overcoming some of the limitations associated with conventional breeding methods (Hasan et al. 2019; Yadav et al. 2012).The integration of molecular markers into breeding programs has revolutionized the genetic improvement of mungbean. MAS allows breeders to select for desirable traits at the seedling stage, significantly reducing the time required to develop new varieties. For example, DNA markers linked to mungbean yellow mosaic virus (MYMV) resistance have been identified, enabling the selection of resistant genotypes in breeding populations (Pataczek et al. 2018). Additionally, research has shown that linkage maps can be utilized to identify quantitative trait loci (QTLs) associated with important agronomic traits, such as seed size and disease resistance (Isemura et al. 2012). The conservation of diverse mungbean germplasm is crucial for future breeding efforts. Gene banks and collections of wild relatives provide a reservoir of genetic diversity that can be tapped into for breeding programs. The identification and characterization of these genetic resources are essential for developing improved varieties that can withstand changing climatic conditions and emerging pests (Isemura et al. 2012; Rahevar et al. 2024). The availability of genomic resources, including complete genome sequences, facilitates the identification of candidate genes for targeted breeding efforts. The traditional breeding of mungbean implies with limitations, one of the primary limitations of traditional breeding in mungbean being the narrow genetic base of the cultivated varieties. Most breeding programs rely on a limited number of trait-specific parental lines, which has resulted in a loss of genetic diversity in farmers' fields (Tripathi et al. 2020). This limited variability restricts the potential for yield increases, especially in the face of emerging biotic stresses such as pests and diseases. The reliance on a small spectrum of mungbean variability has made it challenging to develop varieties that can adapt to changing environmental conditions and resist new threats (Kumar et al. 2022). Many important agronomic traits in mungbean, such as disease resistance and drought tolerance, are controlled by multiple genes, making traditional breeding methods less effective. For instance, developing varieties resistant to powdery mildew through backcross breeding is complicated because the resistance is conferred by multiple genes, which are difficult to identify and pyramid in breeding programs (Papan et al. 2024). This complexity often leads to prolonged breeding cycles and uncertain outcomes (Papan et al. 2024). Furthermore, the availability of molecular markers for mungbean breeding is limited, which hampers the ability to conduct marker-assisted selection (MAS) effectively. Reports indicate that the polymorphic information content of existing SSR markers is low, making them less useful for genetic research and breeding programs (Saini et al. 2023). To overcome the limitations of traditional breeding methods in mungbean, integrating modern biotechnological approaches is essential. However, understanding gene functions enables precise targeting of traits like disease resistance and drought tolerance, simplifying the complexity of multi-gene traits. These advancements reduce labor-intensive processes by accelerating trait selection and introgression, bypassing lengthy breeding cycles. Furthermore, leveraging molecular markers and genomic selection can improve resistance to biotic and abiotic stresses, while tapping into the diversity of landraces and wild relatives. Collectively, these strategies enhance breeding efficiency, seed quality, and adaptability, ensuring mungbean meets future agricultural demands sustainably.

3 Leveraging Whole Genome Sequencing for Mungbean Genetic Enhancement

Whole genome sequencing (WGS) has emerged as a critical resource for the genetic study and improvement of mungbean (Somta et al. 2022). Recent advancements in sequencing technologies have dramatically accelerated mungbean genome research, laying a robust foundation for precise gene editing applications. Next-generation sequencing (NGS) technologies, such as Illumina/Solexa, Roche 454, and Ion Torrent, have enabled rapid, cost-effective, and detailed genomic analyses (Prasad et al. 2024b). These platforms have facilitated the sequencing of nuclear genomes, transcriptomes, and chloroplast genomes of mungbean, generating invaluable resources for genetic improvement (Das et al. 2018; Manjunatha et al. 2023). The first milestone in mungbean genome research was the draft genome sequence of the elite breeding line "VC1973A," developed using Illumina and Roche 454 sequencing (Kang et al. 2014). This draft genome covered approximately 431 Mb, representing 80% of the estimated mungbean genome, and included 2748 scaffolds assembled into 11 pseudochromosomes. It enabled the identification of 22,427 protein-coding genes, of which 18,378 were mapped to pseudochromosomes. The draft sequence also provided insights into genetic variations by identifying millions of SNPs and hundreds of thousands of indels, valuable resources for gene mapping and functional studies. Further advancements in sequencing technology, such as third-generation sequencing, have improved the quality and completeness of mungbean genome assemblies (Amarasinghe et al. 2020). For example, the high-quality reference genome of the mungbean cultivar 'Sulv1' was constructed using a combination of second-generation sequencing, third-generation sequencing (single-molecule real-time sequencing and nanopore sequencing), and Hi-C analysis (Yan et al. 2020). This genome assembly covered 473.7 Mb, or 87.8% of the estimated genome size, and identified 33,924 protein-coding genes with a scaffold N50 of 42.4 Mb (Ha et al. 2021). Similarly, the wild mungbean accession 'TC1966' was sequenced, enabling comparative analyses with cultivated varieties. The improvements in genome assemblies have highlighted significant genomic variations, including presence–absence variations, copy number variations, and chromosomal rearrangements. Comparative analyses of reference genomes, such as VC1973A, Sulv1, and TC1966, reveal substantial genome variability within mungbean, offering unique opportunities to identify and edit genes associated with agronomically important traits like yield, stress tolerance, and disease resistance (Della Coletta et al. 2021). Transcriptomics, another critical aspect of genome research, has also seen a surge due to affordable NGS. Transcriptome sequencing has elucidated genome-wide transcript variation, functional elements, and molecular responses to various biotic and abiotic stresses. These studies have focused on challenges such as drought, chilling, waterlogging, heat, and diseases like fusarium wilt and bruchid infestations, providing target genes for editing to enhance stress resilience. The development of high-quality genome sequences and transcriptomic datasets positions mungbean as a promising candidate for advanced gene editing technologies like CRISPR/Cas9. These genomic resources enable the systematic identification and precise editing of genes, paving the way for the development of improved mungbean cultivars with superior traits. However, further efforts in pangenome construction and sequencing unexplored that portions of the mungbean genome will be crucial for resolving genome variations and maximizing its potential for genetic improvement.

4 Unraveling Gene Function in Mungbean for Precision Gene Editing

Gene function in mungbean plays a crucial role in advancing gene editing opportunities, especially when exploring complex traits such as bruchid resistance, powdery mildew resistance, seed size, and plant architecture. The study of these traits, combined with cutting-edge genomic tools, has revealed important insights into the underlying genetic architecture, facilitating more targeted and effective gene editing strategies.

5 Harnessing CRISPR/Cas9 for Mungbean Improvement

The application of CRISPR/Cas9 technology in mungbean improvement represents a significant advancement in plant biotechnology, enabling precise genetic modifications that can enhance various agronomic traits (Bhowmik et al. 2021). Their ease of transformation, small genome sizes, short life cycles, and available genomic resources make them ideal candidates for CRISPR/Cas9-based studies. One of the promising applications of CRISPR/Cas9 in mungbean is the enhancement of nutritional quality. For instance, efforts to increase the methionine content in mungbean proteins through genetic engineering could significantly improve the nutritional profile of this important legume, particularly for populations that rely heavily on legumes for protein (Mendoza et al. 2001). Additionally, CRISPR/Cas9 can be employed to modify genes associated with stress responses, enabling mungbean to better withstand environmental challenges such as drought and high temperatures (Das and Acharjee 2023). Moreover, the integration of CRISPR/Cas9 with other biotechnological approaches, such as marker-assisted selection and genomic selection, can further accelerate the breeding process for improved mungbean varieties. This integrated approach allows for the identification of beneficial alleles and the development of cultivars that are not only high-yielding but also resilient to biotic and abiotic stresses (Kumar et al. 2022). The use of CRISPR/Cas9 in conjunction with advanced omics technologies can also facilitate a deeper understanding of the genetic basis of important traits, leading to more informed breeding decisions (Gudeta et al. 2024). With continued efforts to develop editing pipelines for elite mungbean varieties, CRISPR/Cas9 technology promises to revolutionize mungbean improvement by enabling precise trait modifications and accelerating the breeding cycle.

5.1 Bruchid Resistance

Bruchid resistance in mungbean is controlled by a dominant locus (Somta et al. 2007), Br, which is associated with antibiosis in cotyledons (Somta et al. 2008). Key genes involved in this resistance includeVrBURP1, VrPGIP1, andVrPGIP2. VrBURP1, encoding a protein with a BURP domain, was identified through map-based cloning as the main resistance gene in Mungbean cultivar 'Jangannogdu.' Additionally, g34458and g39185, which also encode BURP domain proteins, were found near the Br locus on chromosome 5 (Nair 2015; Lin et al. 2016). VrPGIP1 and VrPGIP2 encode polygalacturonase inhibitors (PGIPs), which block the digestive enzyme polygalacturonase in bruchid pests like *Callosobruchus maculatus* (Kaewwongwal et al. 2017; Kaewwongwal et al. 2020). This inhibition prevents the pests from digesting the seed's pectin and cell wall, offering protection. SNPs in VrPGIP2 result in amino acid changes, contributing to resistance in various mungbean cultivars. The polygalacturonase inhibitory activity of both proteins has been confirmed through feeding tests (Zhang et al. 2021).These genes represent potential targets for gene editing in mungbean. By optimizing or introducing favorable alleles in VrPGIP1, VrPGIP2, or VrBURP1 using gene-editing tools like CRISPR-Cas9, researchers can develop mungbean varieties with enhanced resistance to bruchids, improving seed storage and reducing post-harvest losses.

5.2 Powdery Mildew (PM) Resistance

Powdery mildew (PM) resistance in mungbean is influenced by several QTLs and genes, with key loci identified for partial resistance. VrMLO12 is a candidate gene for PM resistance, located at the qPMRUM5-3 locus (Yundaeng et al. 2020). This gene encodes a protein in the MLO (Mildew Locus O) family, which plays a role in plant susceptibility to PM (Kusch and Panstruga 2017). Loss-of-function mutations in MLO genes, especially in the calmodulin-binding domain (CaMBD), result in resistance to the fungus. In mungbean, mutations in VrMLO12 (identified in resistant accessions RUM5 and V4718) cause amino acid changes in the CaMBD and transmembrane domain 6 (TM6), leading to resistance. Additionally, other loci, such as qPMRUM5-2, also contribute to resistance. Gene editing, particularly using tools like CRISPR-Cas9, could target VrMLO12 and other MLO genes, such as VrMLO1, VrMLO2, VrMLO5, VrMLO6, VrMLO11, andVrMLO17, to induce loss-of-function mutations and improve resistance to PM (Rispail and Rubiales 2016). These genes represent promising candidates for enhancing stable, broad-spectrum resistance against powdery mildew in mungbean crops, offering potential for genetically engineered or edited resistant varieties.

5.3 Seed Size

One of the pivotal studies on seed size improvement in mungbean is the work by (Liu et al. 2022), which conducted a comprehensive GWAS to identify quantitative trait loci (QTLs) associated with seed size traits. This research highlighted the genetic variation present in mungbean and identified specific single-nucleotide polymorphisms (SNPs) linked to seed size, providing a foundation for future breeding programs aimed at enhancing seed yield and quality. Similarly, Manjunatha et al. explored the phenological and agronomic traits of mungbean, revealing a high degree of variation in economically important traits, including seed weight and size, which can be targeted for genetic improvement (Manjunatha et al. 2023). Seed size is a polygenic trait with high heritability, and several studies have mapped QTLs for seed weight. In an F2:3 population, (Alam et al. 2014) identified four major QTLs for seed weight: qSDWT1.1, qSDWT6.1, qSDWT8.1, andqSDWT9.1, with qSDWT8.1 showing the largest effect. qSDWT1.1 was found in multiple legume species, while qSDWT6.1 and qSDWT8.1 were identified in mungbean (Alam et al. 2014; Somta et al. 2015). qSDWT8.1 showed the largest effect, making it a prime candidate for further investigation. Recent GWAS analyses have identified 98 quantitative trait nucleotides (QTNs) and 20 QTN-by-environment interactions (QEIs) for seed size traits using advanced statistical methods. Among these, VrEmp24/25 and VrKIX8 emerged as significant candidate genes near stable QTNs, confirmed by RT-qPCR and haplotype analysis (Liu et al. 2022). The prevalent haplotypes of these genes can be utilized in breeding programs for yield improvement. These findings offer a pathway for precision gene editing to optimize seed size and meet industry demands.

5.4 Plant Architecture and Phenology

Plant architecture and phenology traits like stem determinacy, flowering time, and photoperiod response are crucial for mungbean breeding. One significant finding in this area is the identification of the phytochrome E gene (PHYE), which has been pinpointed as a major QTL controlling flowering time in mungbean. This gene plays a critical role in the plant's response to photoperiod, which is particularly important as mungbean is a quantitative short-day plant. The sensitivity of flowering time to photoperiod can significantly affect yield, as it determines the duration of the growing season (Amkul et al. 2024). Furthermore, stem determinacy, controlled by the VrDet1 gene (TFL1 ortholog), is essential for modern cultivars with compact growth and early maturity. Mutations in the VrDet1 promoter region cause the shift from indeterminate to determinate growth, with specific SNPs in regulatory elements crucial for VrDet1 function. Gene editing techniques could be used to manipulate VrDet1 and similar genes for more efficient control of plant architecture (Li et al. 2018). High-resolution mapping has identified Vradi04g00002764 and Vradi04g00002773 on chromosome 4, which exhibit pleiotropic effects on plant height, node number, and synchronous pod maturity, but trade-offs between traits require careful consideration (Ha et al. 2021). Comparative genomics offers opportunities to identify homologs for photoperiod and flowering regulation. These findings suggest that precise gene editing of VrDet1, VrPHYA, and other candidate genes could accelerate mungbean breeding for early flowering, compact plant habit, and higher yield potential.

5.5 Seed Dormancy

Seed dormancy is a critical trait for mungbean domestication, with wild mungbean exhibiting strong dormancy to protect against preharvest sprouting. The gene VrKNAT7–1, located at the Sdwa5.1.1+ locus, regulates physical dormancy by influencing seed coat characteristics, such as cuticle biosynthesis, which prevents germination. Sequence analysis of VrKNAT7–1 in wild and cultivated Mungbean showed polymorphisms linked to dormancy, with wild Mungbean expressing higher levels of VrKNAT7–1 and related genes like VrCYP86A. Gene editing tools could be employed to modify VrKNAT7–1 expression to control seed dormancy, ensuring uniform germination in cultivated Mungbean (Laosatit et al. 2022). Additionally, the Sdwa5.1.2+ locus, containing genes like Vradi07g10480 encoding calmodulin-like protein 1, is involved in the ABA-mediated inhibition of germination, suggesting that gene editing at this locus could regulate seed dormancy. The Sdwa5.3.1+ locus, linked to GA2OX1, a gene controlling gibberellin catabolism, also plays a role in seed dormancy (Isemura et al. 2012). By using gene editing to modify the GA2OX1gene, it may be possible to fine-tune seed dormancy and germination timing. These genetic insights provide potential targets for gene editing to improve seed dormancy control and enhance mungbean yield and quality.

6 Challenges in Mungbean Gene Editing

6.1 Off-Target Effects in Gene Editing of Mungbean

Off-target effects present a significant challenge in the application of gene editing technologies such as CRISPR/Cas9 in mungbean. These effects arise when the gene-editing tool introduces unintended modifications at genomic sites that share sequence similarity with the target site, potentially leading to disruptions in essential genes or regulatory elements (Singer et al. 2021). In mungbean, the relatively small genome size and high sequence homology among gene families heighten the risk of off-target effects. Factors influencing these unintended modifications include sequence similarity, sgRNA design, chromatin accessibility, and the activity of the Cas9 enzyme (Liu et al. 2022). Studies have shown that off-target mutations can compromise plant growth, yield, and resistance traits, making their detection and minimization crucial (Zhang et al. 2020). Detection methods such as whole-genome sequencing (WGS), GUIDE-seq, and in silico prediction tools have been employed to identify off-target sites in gene-edited plants, including legumes (Ahmad et al. 2021; Ullah et al. 2023). However, these methods remain costly and time-consuming, limiting their widespread adoption in mungbean research. Efforts to mitigate off-target effects in mungbean include the use of high-fidelity Cas9 variants (e.g., SpCas9-HF1 and eSpCas9), optimized sgRNA design algorithms, and ribonucleoprotein (RNP) delivery systems to minimize persistent Cas9 activity (Guo et al. 2023). Additionally, employing alternative genome-editing tools such as base editors and prime editors offers a more precise approach with reduced off-target risks (Kantor et al. 2020). Despite these advances, achieving high specificity in mungbean gene editing remains a challenge due to the complexity of its genome and transformation recalcitrance.

7 Overcoming Transformation Recalcitrance in Mungbean Gene Editing

A significant bottleneck in applying genome editing to mungbean is its transformation recalcitrance, a trait shared by most legumes. While some crops are highly amenable to transformation, legumes generally exhibit lower transformation efficiencies. Even among widely cultivated species, only certain cultivars achieve appreciable transformation efficiency (Xu et al. 2022) and mungbean is no exception. Studies have demonstrated its low transformation efficiency, with cotyledonary nodes achieving only 4.2% efficiency (Yadav et al. 2012) and shoot tips reaching a mere 1.49% efficiency (Mekala et al. 2016). These rates fall well below the 15% threshold commonly used to distinguish recalcitrant species. Such low efficiencies increase the labor required for handling explants and reduce the probability of obtaining edited plants (Nivya and Shah 2023).

One of the key challenges lies in the steps of transformation itself: the entry of foreign DNA into the cell and nucleus, successful transgene integration, and regeneration under selection pressure. Despite high GUS, GFP, and YFP reporter gene expression in callus cells indicating efficient DNA uptake and integration, the real limitation arises in the regeneration of transformed cells under selection pressure (Bekalu et al. 2023). This is particularly concerning given the in vitro regeneration potential of mungbean. The application of selection agents (antibiotic selection) during transformation hampers regeneration efficiency, leading to a reduced rate of successful transgene integration and transformed plant regeneration (Nivya and Shah 2023). Inconsistent molecular validation methods further complicate the transformation process. While PCR has been used to confirm transgene integration in the first-generation transgenics, there are often inconsistencies with Southern blot results, and some studies lack crucial junction fragment analysis, which is essential for distinguishing independent transformation events. Furthermore, the inability to rule out *Agrobacterium* contamination using PCR can lead to false-positive results, thus complicating the verification of successful transformation (Jaiwal et al. 2001; Saini et al. 2007; Yadav et al. 2012). Moreover, the inheritance of transgenes in the offspring of primary transformants is often unstable, as no studies confirm stable transgene presence, pointing to the formation of chimeric plants (Mekala et al. 2016). This instability poses a significant challenge in gene editing, where ensuring the transmission of edited loci to progeny is critical for successful applications.

8 Strategies to Overcome Challenges in Gene Editing

8.1 Optimizing Gene Editing Tools for Mungbean

The CRISPR/Cas9 system has revolutionized functional genomics in legumes, offering precise and efficient gene-editing capabilities to address key agricultural challenges in crops such as mungbean. However, successful implementation in mungbean requires careful optimization of various factors, including promoter selection, guide RNA (gRNA) design, and advanced strategies such as tissue-specific editing to overcome the species' inherent transformation and regeneration challenges.

8.2 Promoter Selection and Cell-Type-Specific Gene Editing Strategies

One of the key challenges in gene editing is achieving high transformation efficiency and precise spatial and temporal control over gene expression. Traditional approaches using constitutive promoters, such as the widely used CaMV 35S promoter, often result in pleiotropic effects and unwanted phenotypic changes. Recent studies have demonstrated that species-specific and tissue-specific promoters, can significantly enhance editing efficiency while reducing unintended effects (Zhang et al. 2024). The identification and utilization of cell-type-specific promoters, such as those driving expression in infection sites or nodules, could facilitate more targeted gene edits with minimal off-target consequences.

A significant advancement in genome editing is the development of CRISPR-based tissue-specific knock-out (CRISPR-TSKO) systems, which utilize tissue- or cell-specific promoters to drive Cas9 expression selectively in particular plant tissues (Decaestecker et al. 2019). In mungbean, leveraging tissue-specific promoters such as those linked to nodule development or stress response pathways could allow precise functional studies without compromising overall plant growth and productivity.

8.3 Guide RNA Design and Alternative Gene Editing Approaches

Effective gene editing hinges on the careful design of gRNAs to maximize target specificity and minimize off-target effects. Studies indicate that gRNAs with a GC content of 40–60% and a length of 20 nucleotides generally achieve the highest efficiencies (Jain et al. 2023). In mungbean, bioinformatics tools such as CRISPR-P, CHOPCHOP, and CRISPR RGEN Tools can be utilized to identify optimal target sites with minimal off-target potential (Li et al. 2023a, b). Additionally, designing multiple gRNAs targeting different exons within a gene or using dual-gRNA strategies can enhance knockout efficiency and minimize the likelihood of escape mutations. Advances such as the use of polycistronic-tRNA-gRNA (PTG) systems, which allow multiplexed gene editing, offer a promising strategy for mungbean to target multiple genes simultaneously for complex trait improvement, such as bruchid resistance and abiotic stress tolerance (Kim et al. 2021). The use of high-fidelity Cas9 variants, such as eSpCas9 and Cas9-HF, has been shown to reduce off-target mutations while maintaining robust editing efficiency (Rabinowitz et al. 2024). Moreover, the introduction of Cas12a (Cpf1), which recognizes different protospacer adjacent motifs (PAMs) compared to Cas9, offers an alternative for improving specificity and enabling more flexible targeting within the mungbean genome (Rasheed et al. 2022).

To further refine the gene editing process in mungbean, alternative CRISPR-based techniques such as base editing and prime editing can be employed to introduce targeted point mutations without generating double-strand breaks (Xie et al. 2022). Base editing tools, such as cytosine base editors (CBEs) and adenine base editors (ABEs), have shown potential in creating single-nucleotide polymorphisms associated with stress tolerance and yield improvement in legumes (Baloglu et al. 2022). Prime editing, which allows precise "search-and-replace" editing, could further enhance mungbean breeding by enabling the introduction of specific beneficial alleles with minimal genomic disruption (Lin et al. 2020). So far, prime editing has been successfully implemented in legume protoplasts (Biswas et al. 2022), highlighting the need for further optimization to harness its potential in mungbean improvement.

8.4 Validation, Screening, and Efficiency Improvement Strategies

Despite advancements in editing tools, the validation and screening of edited lines remain crucial to ensure success in mungbean genome editing. Traditional molecular techniques such as PCR and sequencing are often labor-intensive and time-consuming. Recent developments, such as the mutation site-based-specific primers PCR (MSBSP-PCR), offer a rapid and cost-effective method to distinguish between wild-type and homozygous mutants (Guo et al. 2018). Additionally, novel visible markers such as anthocyanins and betacyanins, which enable easy identification of transgenic roots without requiring fluorescence microscopy, could simplify the screening process in mungbean transformation studies (Ruan et al. 2021; Chen et al. 2024).

9 Strategies to Enhance Mungbean Transformation

Improving crop transformation efficiency is essential for advancing genome editing, as discussed by Chen et al. (2022a). In mungbean, transformation efficiency is directly linked to regeneration efficiency, which is influenced by multiple factors, including medium components, phytohormones, explant type, and light conditions. Recent studies highlight the role of molecular signals in organogenesis and embryogenesis, particularly in response to explant wounding, induced plant cell death, and phytohormone interactions (Bidabadi and Jain 2020). Establishing a reproducible whole-plant regeneration system is crucial for developing an effective genetic transformation protocol. Among the commonly used *Agrobacterium* strains for mungbean transformation are EHA105, LBA4404, and C58C1, with others like EHA101 and GV2260 also reported (Aasim et al. 2019). Additionally, *Agrobacterium rhizogenes* has been utilized for anthocyanin regulation in mungbean (Chen et al. 2023). However, mungbean is a poor host for most *Agrobacterium* strains, limiting transformation efficiency. Exploring alternative bacterial species could offer a promising solution. For instance, Cho et al. (2022) achieved the highest transformation efficiency (35%) using *Ochrobactrum haywardense* H1, compared to *Agrobacterium* strains AGL1 (26%) and LBA4404 (12%). Other natural genetic engineers, such as *Rhizobium etli*, *Ochrobactrum haywardense*, and *Ensifer adhaerens*, should also be investigated for their potential in mungbean transformation (Rathore and Mullins 2018).

While *Agrobacterium*-mediated transformation is a cornerstone of plant biotechnology, its low efficiency and genotype dependence remain challenges. Strategies such as modifying the Ti plasmid for more efficient gene delivery, using morphogenic transcription factor (MTF) genes, and employing non-integrating vectors offer promising solutions (Liu et al. 2024), which have significantly improved transformation and regeneration efficiencies (Lowe et al. 2016; Chen et al. 2022b). Given that transformation efficiency is directly tied to regeneration, identifying similar growth-promoting factors for mungbean could enhance its transformation potential.

For example, overexpressing the GOLDEN2 gene, a member of the GARP transcription factor superfamily, has been shown to enhance callus regeneration in rice and maize by activating chloroplast development and influencing phytohormone signaling (Luo et al. 2023). Identifying and utilizing legume-specific homologs of GOLDEN2 could improve mungbean regeneration and, consequently, its transformation efficiency. Incorporating such growth-promoting factors during transformation could help overcome the current limitations in mungbean gene editing.

10 Strategies for Genome Editing Bypassing Transformation

Virus-mediated genome editing presents a promising approach for mungbean (*Vigna radiata*) gene editing, bypassing traditional transformation methods. While viral vectors have been employed for gene silencing (Zaulda et al. 2022; Singh et al. 2022a, b, c), their inability to integrate transgenes is advantageous for genome editing (Nivya and Shah 2023). However, most virus-mediated approaches require *Agrobacterium* to express Cas9 due to its large size (~1368 aa), which limits their applicability. To overcome this limitation, two strategies have been proposed: utilizing smaller nucleases such as CasΦ U, Cas12f1, TnpB, and IscBs, or using viruses with larger cargo capacities, including Potato virus X and Barley yellow striate mosaic virus (He et al. 2022). Since mungbean is a host to various viruses (Akram et al. 2024), these viruses could be explored as genome-editing vectors. The pea early-browning virus and tobacco rattle virus has already been used successfully for CRISPR/Cas9-based editing in *Nicotiana* and *Arabidopsis* (Zhang et al. 2022), suggesting that similar virus vectors could be adapted for mungbean genome editing.

Enhancing heritability in mungbean can be achieved by employing mobile sgRNAs targeting the apical meristem (Ghogare et al. 2021). Further optimization includes codon optimization and species-specific promoters to increase editing efficiency (Baloglu et al. 2022), similar to the chickpea-optimized Cas9 system driven by the *M. truncatula* U6.1 promoter (Gupta et al. 2023). Beyond conventional CRISPR/Cas9, nucleases derived from *Acidaminococcus* and Lachnospiraceae have been effective in transgenic development (Wang et al. 2021) (Singh et al. 2024) and could be tested in mungbean. A proposed "dual-editing" approach can streamline selection by delivering Cas9/sgRNA constructs targeting both a gene of interest (GOI) and a visible marker like phytoene desaturase (PDS) or acetohydroxy acid synthase (AHAS) (Prasad et al. 2024a, b; Wei et al. 2023). This would allow for easy phenotypic selection, as PDS mutants exhibit chlorophyll reduction, while AHAS mutants display herbicide resistance (Siddappa et al. 2023; Galili et al. 2021). This strategy enhances the probability of heritable mutations, enabling the generation of homozygous mutants by the T_2 generation while avoiding transgene integration, which is a limitation in transformation-based methods (Zhang et al. 2022).

Another promising transgene-free genome editing approach is the direct delivery of ribonucleoproteins (RNPs) into mungbean protoplasts. This method involves the introduction of pre-assembled Cas9 protein and guide RNA (gRNA) complexes directly into isolated mungbean protoplasts, eliminating the need for stable transgene integration. RNP-based genome editing has been successfully demonstrated in soybean protoplasts (Subburaj et al. 2022), highlighting its potential for application in mungbean as well. However, RNP-based editing lacks antibiotic selection, requiring labor-intensive PCR screening due to low efficiency (Singh et al. 2022a, b, c). Integrating the "dual-editing" strategy with protoplast methods could improve selection efficiency by introducing sgRNAs directly into protoplasts rather than using viral vectors (Li et al. 2023a, b). Additionally, emerging nanotechnology-based delivery systems, such as carbon nanotubes and mesoporous silicon nanoparticles, offer promising advancements in genome editing by facilitating the efficient transport of RNPs into plant cells (Naik et al. 2022; Kumar et al. 2025). These nanocarriers can improve cellular uptake and enhance editing efficiency while maintaining a transgene-free approach.

11 Conclusion

Mungbean is a crucial legume crop with significant potential to enhance food security and nutritional requirements, due to its high protein content, rapid growth cycle, and nitrogen-fixing properties. Nonetheless, its genetic enhancement has been constrained by the drawbacks of conventional breeding techniques, including a limited genetic pool, intricate multi-gene characteristics, and vulnerability to biotic and abiotic challenges. The emergence of contemporary genomic technologies, especially WGS and CRISPR/Cas9 gene editing, has revolutionary prospects to address these obstacles. WGS has yielded extensive insights into the genetic architecture of mungbean, facilitating the identification of critical genes linked to agronomically significant features such as bruchid resistance, powdery mildew resistance, seed size, and plant architecture. Simultaneously, successful gene editing in model legumes and associated crops such as soybean and cowpea illustrates the potential for applying these innovations to mungbean, offering precise trait improvements and expedited breeding processes. Notwithstanding this potential, considerable obstacles persist, particularly transformation recalcitrance and off-target effects in gene editing. The low transformation efficiency of mungbean, together with challenges in regeneration and unstable transgenic inheritance, requires creative approaches, including tissue-specific promoters, high-fidelity Cas9 variants, and alternate delivery vehicles such as RNPs and viral vectors. Enhancing gRNA design, utilizing base and prime editing, and employing growth-promoting agents such as BBM and WUS can improve editing precision and efficiency. Additionally, virus-mediated and nanotechnology-based techniques offer transgene-free options, avoiding typical transformation bottlenecks and boosting heredity. By merging these cutting-edge tactics with the extensive genomic tools presently accessible, mungbean breeding can be transformed to generate resilient, high-yielding cultivars that suit the demands of a changing agricultural landscape. Continued research and investment in these technologies will be vital to unleashing mungbean's full potential, ensuring its sustainability, and confirming its place in global agriculture.

References

Aasim M, Barpete S, Usta A, Sevinc C (2019) An insight into Agrobacterium tumafaciens-mediated genetic transformation studies in Mungbean (*Vigna radiata* L. Wilczek). J Global Innov Agric Sci 7:47–51

Ahmad A, Munawar N, Khan Z, Qusmani AT, Khan SH, Jamil A, Ashraf S, Ghouri MZ, Aslam S, Mubarik MS, Munir A, Sultan Q, Abd-Elsalam KA, Qari SH (2021) An outlook on global regulatory landscape for genome-edited crops. Int J Mol Sci 22:11753. https://doi.org/10.3390/ijms222111753

Akram M, Kamaal N, Pratap A, Kumar D, Muin A, Sabale PR, Aidbhavi R, Sunani SK, Rathore M, Gupta S, Singh NP, Dey N, Dixit GP, Nair RM (2024) Exploring distribution and genomic diversity of begomoviruses associated with yellow mosaic disease of legume crops from India, highlighting the dominance of Mungbean yellow mosaic India virus. Front Microbiol 15. https://doi.org/10.3389/fmicb.2024.1451986

Alam AKMM, Somta P, Muktadir MA, Srinives P (2014) Quantitative trait loci associated with seed weight in Mungbean (*Vigna radiata* (L.) Wilczek). Agric Nat Resour 48:197–204

Amarasinghe SL, Su S, Dong X, Zappia L, Ritchie ME, Gouil Q (2020) Opportunities and challenges in long-read sequencing data analysis. Genome Biol 21:30. https://doi.org/10.1186/s13059-020-1935-5

Amkul K, Laosatit K, Lin Y, Yimram T, Chen J, Yuan X, Chen X, Somta P (2024) Narrowing down a major QTL region reveals phytochrome E (PHYE) as the candidate gene controlling flowering time in Mungbean (*Vigna radiata*). Breed Sci 74:83–92. https://doi.org/10.1270/jsbbs.23036

Aneja B, Yadav NR, Yadav RC, Kumar R (2013) Sequence related amplified polymorphism (SRAP) analysis for genetic diversity and micronutrient content among gene pools in Mungbean (*Vigna radiata* (L.) Wilczek). Physiol Mol Biol Plants 19:399–407. https://doi.org/10.1007/s12298-013-0177-3

Aziz-ur-Rehman M, Kaukab S, Saeed S, Aqeel M, Riasat G, Rafiq CM (2019) Prospects of Mungbean as an additional crop in rice wheat system of Punjab Pakistan. Univers J Agric Res 7:136–141

Baloglu MC, Celik Altunoglu Y, Baloglu P, Yildiz AB, Türkölmez N, Özden Çiftçi Y (2022) Gene-editing technologies and applications in legumes: Progress, evolution, and future prospects. Front Genet 13:859437. https://doi.org/10.3389/fgene.2022.859437

Bekalu ZE, Panting M, Bæksted Holme I, Brinch-Pedersen H (2023) Opportunities and challenges of in vitro tissue culture systems in the era of crop genome editing. Int J Mol Sci 24:11920. https://doi.org/10.3390/ijms241511920

Bhowmik P, Konkin D, Polowick P, Hodgins CL, Subedi M, Xiang D, Yu B, Patterson N, Rajagopalan N, Babic V, Ro DK, Tar'an B, Bandara M, Smyth SJ, Cui Y, Kagale S (2021) CRISPR/Cas9 gene editing in legume crops: opportunities and challenges. Legume Sci 3:e96. https://doi.org/10.1002/leg3.96

Bidabadi SS, Jain SM (2020) Cellular, molecular, and physiological aspects of in vitro plant regeneration. Plants 9(6):702. https://doi.org/10.3390/plants9060702

Biswas S, Bridgeland A, Irum S, Thomson MJ, Septiningsih EM (2022) Optimization of prime editing in rice, peanut, chickpea, and cowpea protoplasts by restoration of GFP activity. Int J Mol Sci 23(7):9809. https://doi.org/10.3390/ijms23179809

Chen Z, Debernardi JM, Dubcovsky J, Gallavotti A (2022a) Recent advances in crop transformation technologies. Nat Plants 8:1343–1351. https://doi.org/10.1038/s41477-022-01295-8

Chen Z, Debernardi JM, Dubcovsky J, Gallavotti A (2022b) The combination of morphogenic regulators BABY BOOM and GRF-GIF improves maize transformation efficiency bioRxiv, 2022-09. https://doi.org/10.1101/2022.09.02.506370

Chen J, Zhan J, Wang H, Zhao Y, Zhang D, Chen X, Su N, Cui J (2023) VrMYB90 functions synergistically with VrbHLHA and VrMYB3 to regulate anthocyanin biosynthesis in Mungbean. Plant Cell Physiol 64:221–233. https://doi.org/10.1093/pcp/pcac160

Chen L, Cai Y, Liu X, Yao W, Wu S, Hou W (2024) The RUBY reporter for visual selection in soybean genome editing. A Biotech 5:209–213. https://doi.org/10.1007/s42994-024-00148-6

Cho HJ, Moy Y, Rudnick NA, Klein TM, Yin J, Bolar J, Hendrick C, Beatty M, Castañeda L, Kinney AJ, Jones TJ (2022) Development of an efficient marker-free soybean transformation method using the novel bacterium Ochrobactrum haywardense H1. Plant Biotechnol J 20(5):977–990. https://doi.org/10.1111/pbi.13777

Dahiya PK, Linnemann AR, Van Boekel M, Khetarpaul N, Grewal RB, Nout MJR (2015) Mungbean: technological and nutritional potential. Crit Rev Plant Sci 55:670–688. https://doi.org/10.1080/10408398.2012.671202

Das D, Acharjee S (2023) Application of CRISPR/Cas-mediated genome editing techniques in leguminous crops. In: A roadmap for plant genome editing. Springer, Cham, pp 247–258

Das A, Singh P, Kushwah NS, Thakur S, Rathore M, Pratap A, Singh.NP (2018) Genetic and genomic approaches for improvement in Mungbean (*Vigna radiata* L.). In: Wani SH, Jain M (eds) Pulse improvement: physiological, molecular and genetic perspectives. Springer, Cham, pp 175–189. https://doi.org/10.1007/978-3-030-01743-9_8

Decaestecker W, Buono RA, Pfeiffer ML, Vangheluwe N, Jourquin J, Karimi M, Van Isterdael G, Beeckman T, Nowack MK, Jacobs TB (2019) CRISPR-TSKO: a technique for efficient mutagenesis in specific cell types, tissues, or organs in Arabidopsis. Plant Cell 31:2868–2887. https://doi.org/10.1105/tpc.19.00454

Della Coletta R, Qiu Y, Ou S, Hufford MB, Hirsch CN (2021) How the pan-genome is changing crop genomics and improvement. Genome Biol 22:3. https://doi.org/10.1186/s13059-020-02224-8

Diatta AA, Abaye O, Battaglia ML, JFDC L, Seleiman M, Babur E, Thomason WE (2024) Mungbean (*Vigna radiata* (L.) Wilczek) and its potential for crop diversification and sustainable food production in Sub-Saharan Africa: a review. Technology in Agronomy 4:e031. https://doi.org/10.48130/tia-0024-0030

Galili S, Hershenhorn J, Edelman M, Sobolev V, Smirnov E, Amir-Segev O, Bellalou A, Dor E (2021) Novel mutation in the acetohydroxyacid synthase (AHAS) gene confers imidazolinone resistance in chickpea *Cicer arietinum* L. Plants 10:2791. https://doi.org/10.3390/plants10122791

Ghogare R, Ludwig Y, Bueno GM, Slamet-Loedin IH, Dhingra A (2021) Genome editing reagent delivery in plants. Transgenic Res 30:321–335. https://doi.org/10.1007/s11248-021-00239-w

Gudeta TB, Keneni G, Figlan S (2024) Progress in the use of combined omics for Mungbean breeding improvement and its potential in promoting resistance against Cercospora leaf spot. Int J Plant Biol 15:482–504. https://doi.org/10.3390/ijpb15020037

Guo J, Li K, Jin L, Xu R, Miao K, Yang F, Qi C, Zhang L, Botella JR, Wang R, Miao Y (2018) A simple and cost-effective method for screening of CRISPR/Cas9-induced homozygous/biallelic mutants. Plant Methods 14:40. https://doi.org/10.1186/s13007-018-0305-8

Guo C, Ma X, Gao F, Guo Y (2023) Off-target effects in CRISPR/Cas9 gene editing. Front Bioeng Biotechnol 11:1142095. https://doi.org/10.3389/fbioe.2023.1143157

Gupta SK, Vishwakarma NK, Malakar P, Vanspati P, Sharma NK, Chattopadhyay D (2023) Development of an Agrobacterium-delivered codon-optimized CRISPR/Cas9 system for chickpea genome editing. Protoplasma 260:1437–1451. https://doi.org/10.1007/s00709-023-01856-4

Gwag JG, Dixit A, Park YJ, Ma KH, Kwon SJ, Cho GT, Lee GA, Lee SY, Kang HK, Lee SH (2010) Assessment of genetic diversity and population structure in Mungbean. Genes Genome 32:299–308. https://doi.org/10.1007/s13258-010-0014-9

Ha J, Satyawan D, Jeong H, Lee E, Cho KH, Kim MY, Lee SH (2021) A near-complete genome sequence of Mungbean (*Vigna radiata* L.) provides key insights into the modern breeding program. Plant Genome 14:e20121. https://doi.org/10.1002/tpg2.20121

Hasan MM, Bhajan SK, Hoque MI, Sarker RH, Islam MN (2019) In planta genetic transformation of Mungbean (*Vigna radiata* (L.) Wilczek) with marker gene. Plant Tissue Cult Biotechnol 29:245–255

He Y, Mudgett M, Zhao Y (2022) Advances in gene editing without residual transgenes in plants. Plant Physiol 188:1757–1768. https://doi.org/10.1093/plphys/kiab574

Isemura T, Kaga A, Tabata S, Somta P, Srinives P, Shimizu T, Jo U, Vaughan DA, Tomooka N (2012) Construction of a genetic linkage map and genetic analysis of domestication related traits in Mungbean (*Vigna radiata*). Theor Appl Genet 125:1327–1337

Jain D, Jones L, Roy S (2023) Gene editing to improve legume-rhizobia symbiosis in a changing climate. Curr Opin Plant Biol 71:102324. https://doi.org/10.1016/j.pbi.2022.102324

Jaiwal PK, Kumari R, Ignacimuthu S, Potrykus I, Sautter C (2001) Agrobacterium tumefaciens-mediated genetic transformation of Mungbean (*Vigna radiata* L. Wilczek) – a recalcitrant grain legume. Plant Sci 161:239–247. https://doi.org/10.1016/S0168-9452(01)00352-1

Kaewwongwal A, Chen J, Somta P, Kongjaimun A, Yimram T, Chen X, Srinives P (2017) Novel alleles of two tightly linked genes encoding polygalacturonase-inhibiting proteins (VrPGIP1 and VrPGIP2) associated with the Br locus that confer bruchid (Callosobruchus spp.) resistance to Mungbean (*Vigna radiata*) accession V2709. Front Plant Sci 8:1692. https://doi.org/10.3389/fpls.2017.01692

Kaewwongwal A, Liu C, Somta P, Chen J, Tian J, Yuan X, Chen X (2020) A second VrPGIP1 allele is associated with bruchid resistance (Callosobruchus spp.) in wild Mungbean (*Vigna radiata* var. sublobata) accession ACC41. Mol Gen Genomics 295:275–286. https://doi.org/10.1007/s00438-019-01619-y

Kang YJ, Kim SK, Kim MY, Lestari P, Kim KH, Ha BK, Jun TH, Hwang WJ, Lee T, Lee J, Shim S, Yoon MY, Jang YE, Han KS, Taeprayoon P, Yoon N, Somta P, Tanya P, Kim KS, Lee SH (2014) Genome sequence of Mungbean and insights into evolution within *Vigna* species. Nat Commun 5:5443. https://doi.org/10.1038/ncomms6443

Kantor A, McClements ME, MacLaren RE (2020) CRISPR-Cas9 DNA base-editing and prime-editing. Int J Mol Sci 21:6240. https://doi.org/10.3390/ijms21176240

Kim WN, Kim HJ, Chung YS, Kim HU (2021) Construction of multiple guide RNAs in CRISPR/Cas9 vector using stepwise or simultaneous Golden Gate cloning: case study for targeting the FAD2 and FATB multigene in soybean. Plants 10:2542. https://doi.org/10.3390/plants10112542

Kumar AY, Seema P, Kumar RR, Kushwaha A, Rashid MH, Tarannum N, Chakraborty S (2022) Mungbean (*Vigna radiata* (L.) R. Wilczek): progress in breeding and future challenges. Int J Plant Soil Sci 34:50–59

Kumar P, Rajput VD, Singh AK, Agrawal S, Das R, Minkina T, Shukla PK, Wong MH, Kaushik A, Albukhaty S, Tiwari KN, Mishra SK (2025) Nano-assisted delivery tools for plant genetic engineering: a review on recent developments. Environ Sci Pollut Res 32:469–484. https://doi.org/10.1007/s11356-024-35806-1

Kusch S, Panstruga R (2017) mlo-based resistance: an apparently universal "weapon" to defeat powdery mildew disease. Mol Plant-Microbe Interact 30:179–189. https://doi.org/10.1094/MPMI-12-16-0255-CR

Laosatit K, Amkul K, Yimram T, Chen J, Lin Y, Yuan X, Wang L, Chen X, Somta P (2022) A class II KNOX gene, KNAT7-1, regulates physical seed dormancy in Mungbean (*Vigna radiata* (L.) Wilczek). Front Plant Sci 13:852373. https://doi.org/10.3389/fpls.2022.852373

Li S, Ding Y, Zhang D, Wang X, Tang X, Dai D, Jin H, Lee SH, Cai C, Ma J (2018) Parallel domestication with a broad mutational spectrum of determinate stem growth habit in leguminous crops. Plant J 96:761–771. https://doi.org/10.1111/tpj.14066

Li C, Chu W, Gill RA, Sang S, Shi Y, Hu X, Yang Y, Zaman QU, Zhang B (2023a) Computational tools and resources for CRISPR/Cas genome editing. Genomics Proteomics Bioinformatics 21:108–126. https://doi.org/10.1016/j.gpb.2022.02.006

Li Y, Huang C, Liu Y, Zeng J, Yu H, Tong Z, Yuan X, Sui X, Fang D, Xiao B, Zhao S, Yuan C (2023b) CRISPR/Cas9-mediated seamless gene replacement in protoplasts expands the resistance spectrum to TMV-U1 strain in regenerated Nicotiana tabacum. Plant Biotechnol J 21:2641–2653. https://doi.org/10.1111/pbi.14159

Liang J, He Z, Shi W (2020) Cotton/Mungbean intercropping improves crop productivity, water use efficiency, nitrogen uptake, and economic benefits in the arid area of Northwest China. Agric Water Manag 240:106277. https://doi.org/10.1016/j.agwat.2020.106277

Lin WJ, Ko CY, Liu MS, Kuo CY, Wu DC, Chen CY, Schafleitner R, Chen LFO, Lo HF (2016) Transcriptomic and proteomic research to explore bruchid-resistant genes in Mungbean isogenic lines. J Agric Food Chem 64:6648–6658. https://doi.org/10.1021/acs.jafc.6b03015

Lin Q, Zong Y, Xue C, Wang S, Jin S, Zhu Z, Wang Y, Anzalone AV, Raguram A, Doman JL, Liu DR, Gao C (2020) Prime genome editing in rice and wheat. Nat Biotechnol 38:582–585. https://doi.org/10.1038/s41587-020-0455-x

Liu J, Lin Y, Chen J, Yan Q, Xue C, Wu R, Chen X, Yuan X (2022) Genome-wide association studies provide genetic insights into natural variation of seed-size-related traits in Mungbean. Front Plant Sci 13:997988. https://doi.org/10.3389/fpls.2022.997988

Liu S, Wang K, Geng S, Hossain M, Ye X, Li A, Mao L, Kogel KH (2024) Enemies at peace: recent progress in *agrobacterium*-mediated cereal transformation. Crop J 12:321–329. https://doi.org/10.1016/j.cj.2023.12.009

Lowe K, Wu E, Wang N, Hoerster G, Hastings C, Cho MJ, Scelonge C, Lenderts B, Chamberlin M, Cushatt J, Wang L, Ryan L, Khan T, Chow-Yiu J, Hua W, Yu M, Banh J, Bao Z, Brink K, Gordon-Kamm W (2016) Morphogenic regulators Baby boom and Wuschel improve monocot transformation. Plant Cell 28:1998–2015. https://doi.org/10.1105/tpc.16.00124

Luo W, Tan J, Li T, Feng Z, Ding Z, Xie X, Chen Y, Chen L, Liu YG, Zhu Q, Guo J (2023) Overexpression of maize GOLDEN2 in rice and maize calli improves regeneration by activating chloroplast development. Sci China Life Sci 66:340–349. https://doi.org/10.1007/s11427-022-2149-2

Manjunatha PB, Aski MS, Mishra GP, Gupta S, Devate NB, Singh A, Bansal R, Kumar S, Nair RM, Dikshit HK (2023) Genome-wide association studies for phenological and agronomic traits in Mungbean (*Vigna radiata* L. Wilczek). Front Plant Sci 14:1209288. https://doi.org/10.3389/fpls.2023.1209288

Mekala GK, Juturu VN, Mallikarjuna G, Kirti PB, Yadav SK (2016) Optimization of Agrobacterium-mediated genetic transformation of shoot tip explants of green gram (*Vigna radiata* (L.) Wilczek). Plant Cell Tissue Organ Cult 127:651–663. https://doi.org/10.1007/s11240-016-1085-3

Mendoza EMT, Adachi M, Bernardo AEN, Utsumi S (2001) Mungbean [*Vigna radiata* (L.) Wilczek] globulins: purification and characterization. J Agric Food Chem 49:1552–1558. https://doi.org/10.1021/jf001041h

Naik BJ, Shimoga G, Kim SC, Manjulatha M, Subramanyam Reddy C, Palem RR, Kumar M, Kim SY, Lee SH (2022) CRISPR/Cas9 and nanotechnology pertinence in agricultural crop refinement. Front Plant Sci 13:843575. https://doi.org/10.3389/fpls.2022.843575

Nair R (2015) The AVRDC–The World Vegetable Center Mungbean (*Vigna radiata*) core and mini core collections. Genet Resour Crop Evol 62:1129–1145. https://doi.org/10.1186/s12864-015-1556-7

Nair RM, Yang R, Easdown WJ, Thavarajah D, Thavarajah P, Hughes J d'A, Keatinge JDH (2013) Biofortification of Mungbean (*Vigna radiata*) as a whole food to enhance human health. J Sci Food Agric 93:1805–1813. https://doi.org/10.1002/jsfa.6110

Nair R, Schafleitner R, Easdown W, Ebert A, Hanson P, d'Arros HJ, Keatinge JDH (2014) Legume improvement program at AVRDC-The World Vegetable Center: impact and future prospects. Field Vegetable Crop Res 51:9–20

Nair RM, Pandey AK, War AR, Hanumantharao B, Shwe T, Alam A, Pratap A, Malik SR, Karimi R, Mbeyagala EK (2019) Biotic and abiotic constraints in Mungbean production progress in genetic improvement. Front Plant Sci 10:1340. https://doi.org/10.3389/fpls.2019.01340

Nivya VM, Shah JM (2023) Recalcitrance to transformation, a hindrance for genome editing of legumes. Front Genome Edit 5:1247815. https://doi.org/10.3389/fgeed.2023.1247815

Papan P, Chueakhunthod W, Khairum A, Siwapitakpong K, Chaiyapan C, Inthaisong S, Jinagool W, Tharapreuksapong A, Masari A, Kaewkasi C (2024) Marker-assisted gene pyramiding for powdery mildew resistance in Thai Mungbean variety SUT1 by backcross breeding. Plant Mol Biol Report 42:1–13. https://doi.org/10.1007/s11105-024-01445-6

Pataczek L, Zahir ZA, Ahmad M, Rani S, Nair R, Schafleitner R, Cadisch G, Hilger T (2018) Beans with benefits the role of Mungbean (*Vigna radiata)* in a changing environment. Am J Plant Sci 9:1577–1591. https://doi.org/10.4236/ajps.2018.97115

Prasad K, Gadeela H, Bommineni PR, Reddy PS, Tyagi W, Yogendra K (2024a) CRISPR/Cas9-mediated mutagenesis of phytoene desaturase in pigeonpea and groundnut. Funct Integr Genomics 24:57. https://doi.org/10.1007/s10142-024-01336-9

Prasad SS, Singh A, Patil K, Ramteke PW, Kumar N, Shukla PK, Yashwanth V, Aakanksha K, Srija MSL (2024b) Next-generation sequencing (NGS) application in plant breeding. In: Al-Khayri JM, Ingle KP, Jain SM, Penna S (eds) Plant molecular breeding in genomics era: concepts and tools. Springer, pp 149–171. https://doi.org/10.1007/978-3-031-68586-6_6

Pratap A, Das A, Kumar S, Gupta S (2021) Current perspectives on introgression breeding in food legumes. Front Plant Sci 11:589189. https://doi.org/10.3389/fpls.2020.589189

Rabinowitz R, Shor O, Zerbib J, Herman S, Zelikson N, Madiwale S, Yom-Tov N, Ben-David U, Benninger F, Offen D (2024) Computationally engineered CRISPR-SpyCas9 high-fidelity variants with improved specificity and reduced non-specific DNA damage. bioRxiv 2023-04. https://doi.org/10.1101/2023.04.11.536265

Rahevar PM, Chauhan RM, Patel PT, Patel MP, Bhadauria HS, Solanki SD, Viradiya YA (2024) Exploring population genetics for mutant Mungbean [*Vigna radiata* (L.) R. Wilczek] lines: insights from augmented block design. Agron J 116:2435–2445. https://doi.org/10.1002/agj2.21637

Rasheed A, Barqawi AA, Mahmood A, Nawaz M, Shah AN, Bay DH, Alahdal MA, Hassan MU, Qari SH (2022) CRISPR/Cas9 is a powerful tool for precise genome editing of legume crops: a review. Mol Biol Rep 49:5595–5609. https://doi.org/10.1007/s11033-022-07529-4

Rathore DS, Mullins E (2018) Alternative non-agrobacterium-based methods for plant transformation. Annu Plant Rev Online, In, pp 891–908. https://doi.org/10.1002/9781119312994.apr0659

Reflinur R, Lestari P, Lee SH (2016) The potential use of SSR markers to support the morphological identification of Indonesian Mungbean varieties. Indonesian J Agric Sci 17:65–74. https://doi.org/10.21082/ijas.v17n2.2016.p65-74

Rispail N, Rubiales D (2016) Genome-wide identification and comparison of legume MLO gene family. Sci Rep 6:32673. https://doi.org/10.1038/srep32673

Ruan Y, Chen K, Su Y, Jiang S, Xu PMJD (2021) A root tip-specific expressing anthocyanin marker for direct identification of transgenic tissues by the naked eye in symbiotic studies. Plants 10:605. https://doi.org/10.3390/plants10030605

Saini R, Singh RP, Jaiwal PK (2007) Agrobacterium tumefaciens mediated transfer of Phaseolus vulgaris α-amylase inhibitor-1 gene into Mungbean *Vigna radiata* (L.) Wilczek using bar as selectable marker. Plant Cell Rep 26:187–198. https://doi.org/10.1007/s00299-006-0224-4

Saini P, Sirari A, Gnanesh BN, Mandahal KS, Ludhar NK, Nagpal S, Patel SAH, Akhatar J, Saini P, Pratap A (2023) Assessment of simple sequence repeat (SSR) markers derived from whole-genome sequence (WGS) of Mungbean (*Vigna radiata* L. Wilczek): cross-species transferability and population genetic studies in *Vigna* species. Horticulturae 10:34. https://doi.org/10.3390/horticulturae10010034

Siddappa S, Sharma N, Salaria N, Thakur K, Pathania S, Singh B, Sharma H, Sood S, Bhardwaj V, Thakur AK, Mangal V, Kumar V, Muruthachallam R, Singh K, Tuli R (2023) CRISPR/Cas9-mediated editing of phytoene desaturase (PDS) gene in an important staple crop, potato. 3 Biotech 13:129. https://doi.org/10.1007/s13205-023-03543-w

Singer SD, Laurie JD, Bilichak A, Kumar S, Singh J (2021) Genetic variation and unintended risk in the context of old and new breeding techniques. Crit Rev Plant Sci 40:68–108. https://doi.org/10.1080/07352689.2021.1883826

Singh V, Yadav NR, Singh J (2017) Role of genomic tools for Mungbean [*Vigna radiata* (L.) Wilczek] improvement. Legum Res 40:601–608. https://doi.org/10.18805/lr.v0i0.8406

Singh R, Kaur N, Praba UP, Kaur G, Tanin MJ, Kumar P, Neelam K, Sandhu JS, Vikal Y (2022a) A prospective review on selectable marker-free genome engineered rice: past, present and future scientific realm. Front Genet 13. https://doi.org/10.3389/fgene.2022.882836

Singh V, Lone RA, Kumar V, Mohanty CS (2022b) Reducing the biosynthesis of condensed tannin in winged bean (*Psophocarpus tetragonolobus* (L) DC) by virus-induced gene silencing of anthocyanidin synthase (ANS) gene. 3 Biotech 13:16. https://doi.org/10.1007/s13205-022-03435-5

Singh P, Pandey B, Pratap A, Gyaneshwari U, Nair RM, Mishra AK, Singh CM (2022c) Genetic and genomics resources of cross-species *Vigna* gene pools for improving biotic stress resistance in Mungbean (*Vigna radiata* L. Wilczek). Agronomy 12:3000. https://doi.org/10.3390/agronomy12123000

Singh S, Mishra V, Riyazuddin R, Chugh V, Upadhyay SK (2024) CRISPR/Cas9-mediated genome editing for trait improvement and stress tolerance in Leguminosae (legume family). Plant Breed. https://doi.org/10.1111/pbr.13234

Sompong U, Somta P, Raboy V, Srinives P (2012) Mapping of quantitative trait loci for phytic acid and phosphorus contents in seed and seedling of Mungbean (*Vigna radiata* (L.) Wilczek). Breed Sci 62:87–92. https://doi.org/10.1270/jsbbs.62.87

Somta P, Ammaranan C, Ooi PA-C, Srinives P (2007) Inheritance of seed resistance to bruchids in cultivated Mungbean (*Vigna radiata* L. Wilczek). Euphytica 155:47–55. https://doi.org/10.1007/s10681-006-9299-9

Somta C, Somta P, Tomooka N, Ooi PA-C, Vaughan DA, Srinives P (2008) Characterization of new sources of Mungbean (*Vigna radiata* (L.) Wilczek) resistance to bruchids, Callosobruchus spp. (Coleoptera: Bruchidae). J Stored Prod Res 44:316–321. https://doi.org/10.1016/j.jspr.2008.04.002

Somta P, Chankaew S, Kongjaimun A, Srinives P (2015) QTLs controlling seed weight and days to flowering in Mungbean [*Vigna radiata* (L.) Wilczek], their conservation in azuki bean [*V. angularis* (Ohwi) Ohwi & Ohashi] and rice bean [*V. umbellata* (Thunb.) Ohwi & Ohashi]. Agrivita 37:172–179. https://doi.org/10.17503/Agrivita.V37i2.578

Somta P, Laosatit K, Yuan X, Chen X (2022) Thirty years of Mungbean genome research: where do we stand and what have we learned? Front. Plant Sci 13:944721. https://doi.org/10.3389/fpls.2022.944721

Subburaj S, Zanatta CB, Nunn JAL, Hoepers AM, Nodari RO, Agapito-Tenfen SZ (2022) A DNA-free editing platform for genetic screens in soybean via CRISPR/Cas9 ribonucleoprotein delivery. Front Plant Sci 13:939997. https://doi.org/10.3389/fpls.2022.939997

Tripathi K, Meena SK, Panwar BS, Lal H, Rana JC, Singh K (2020) Understanding genetic variability in the Mungbean (*Vigna radiata* L.) genepool. Ann Appl Biol 177(3):346–357

Ullah Z, Iqbal J, Abbasi BA, Ijaz S, Sharifi-Rad J, Yaseen T, Uddin S, Mahmood T (2023) How genome editing can be helpful in the biofortification of legumes. In: Nadeem MA, Baloch FS, Fiaz S, Aasim M, Habyarimana E, Sönmez O, Zencirci N (eds) Legumes biofortification. Springer, pp 207–232. https://doi.org/10.1007/978-3-031-33957-8_8

Wang T, Zhang C, Zhang H, Zhu H (2021) CRISPR/Cas9-mediated gene editing revolutionizes the improvement of horticulture food crops. J Agric Food Chem 69:13260–13269. https://doi.org/10.1021/acs.jafc.1c00104

Wei T, Jiang L, You X, Ma P, Xi Z, Wang NN (2023) Generation of herbicide-resistant soybean by base editing. Biology 12:741. https://doi.org/10.3390/biology12050741

Xie Y, Haq SIU, Jiang X, Zheng D, Feng N, Wang W, He JS, Qiu QS (2022) Plant genome editing: CRISPR, base editing, prime editing, and beyond. Grassland Res 1:234–243. https://doi.org/10.1002/glr2.12034

Xu H, Guo Y, Qiu L, Ran Y (2022) Progress in soybean genetic transformation over the last decade. Front Plant Sci 13:900318. https://doi.org/10.3389/fpls.2022.900318

Yadav SK, Katikala S, Yellisetty V, Kannepalle A, Narayana JL, Maddi V, Mandapaka M, Shanker AK, Bandi V, Bharadwaja KP (2012) Optimization of agrobacterium mediated genetic transformation of cotyledonary node explants of *Vigna radiata*. Springer Plus 1:1–8. https://doi.org/10.1186/2193-1801-1-59

Yan Q, Wang Q, Xuzhen C, Wang L, Somta P, Xue C, Chen J, Wu R, Lin Y, Yuan X (2020) High-quality genome assembly, annotation and evolutionary analysis of the Mungbean (*Vigna radiata*) genome. Authorea Preprints. https://doi.org/10.22541/au.160587196.63922177/v1

Yi-Shen Z, Shuai S, FitzGerald R (2018) Mungbean proteins and peptides: nutritional, functional and bioactive properties. Food Nutr Res 62:10–29219

Yundaeng C, Somta P, Chen J, Yuan X, Chankaew S, Srinives P, Chen X (2020) Candidate gene mapping reveals VrMLO12 (MLO Clade II) is associated with powdery mildew resistance in Mungbean (*Vigna radiata* [L.] Wilczek). Plant Sci 298:110594. https://doi.org/10.1016/j.plantsci.2020.110594

Zaulda FA, Yang SH, Han J, Mlotshwa S, Dorrance A, Qu F (2022) A cowpea severe mosaic virus-based vector simplifies virus-induced gene silencing and foreign protein expression in soybean. Plant Methods 18:116. https://doi.org/10.1186/s13007-022-00950-7

Zhang D, Hussain A, Manghwar H, Xie K, Xie S, Zhao S, Larkin RM, Qing P, Jin S, Ding F (2020) Genome editing with the CRISPR-Cas system: an art, ethics and global regulatory perspective. Plant Biotechnol J 18:1651–1669. https://doi.org/10.1111/pbi.13383

Zhang Q, Yan Q, Yuan X, Lin Y, Chen J, Wu R, Xue C, Zhu Y, Chen X (2021) Two polygalacturonase-inhibiting proteins (VrPGIP) of *Vigna radiata* confer resistance to bruchids (Callosobruchus spp.). J Plant Physiol 258–259:153376. https://doi.org/10.1016/j.jplph.2021.153376

Zhang C, Liu S, Li X, Zhang R, Li J (2022) Virus-induced gene editing and its applications in plants. Int J Mol Sci 23:18202. https://doi.org/10.3390/ijms231810202

Zhang C-X, Li R-J, Baude L, Reinhardt D, Xie Z-P, Staehelin C (2024) CRISPR/Cas9-mediated generation of mutant lines in Medicago truncatula indicates a symbiotic role of MtLYK10 during nodule formation. Biology 13:1. https://doi.org/10.3390/biology13010053

LORC: Legume Omics Resources Center

Shen Wang, Yanjun Zhang, Xueqing Wang, Jianxiang Peng, Yan Wang, Changyou Liu, Zhixiao Zhang, Qiuzhu Su, Sivakumar Paramasivam, Jing Tian, and Baojie Fan

Abstract

Mungbean (*Vigna radiata* L.) seeds and sprouts are rich in nutrients and have important human health benefits and medicinal uses. However, understanding the genetic variation, population structure, and genes associated with important traits in mungbean has proven to be challenging. With the rapid development of sequencing technology, genetic data for an increasing number of legume crops are accessible for gene mining. In this study, we first integrated multi-omics datasets for selected legume crops, including genomic data for 217 mungbean accessions, transcriptomic data for five legume species, proteomic data for Mungbean 'Jilv 7,' and phenotypic data for five legume species. The Legume Omics Resources Center (LORC; https://njh.biobin.com.cn/cgi-bin/legume/

Shen Wang, Yanjun Zhang and Xueqing Wang contributed equally with all other contributors.

S. Wang (✉) · Y. Wang · C. Liu · Z. Zhang · Q. Su · J. Tian · B. Fan
Institute of Cereal and Oil Crops, Hebei Academy of Agricultural and Forestry Sciences / Hebei Laboratory of Crop Genetic and Breeding, Shijiazhuang, Hebei, China

Y. Zhang
Department of Resource and Environmental Engineering, Xingtai Polytechnic College, Xingtai, Hebei, China

X. Wang
Hebei Academy of Agricultural and Forestry Sciences, Shijiazhuang, Hebei, China

J. Peng
Biobin Data Science, Changsha, China

S. Paramasivam
Dr. M.S. Swaminathan Agricultural College and Research Institute, Tamil Nadu Agricultural University, Eachangkottai, Thanjavur, Tamil Nadu, India

S. Paramasivam et al. (eds.), *Mungbean*,
https://doi.org/10.1007/978-981-95-7637-1_11

Display.cgi?page=/home) public database was constructed. LORC provides multiple search and visualization functions, which include genomic sequences, genome structure, gene annotations, single-nucleotide polymorphism markers, presence/absence variation information, and gene expression data. Overall, LORC represents a rich genetic resource and incorporates easy-to-use analytical tools, and lays a sound foundation for further analysis of the genetic diversity and breeding of legume crops.

Keywords

Gene Annotations · Genome Structure · Mungbean · LORC

1 Introduction

Mungbean (*Vigna radiata* L.) is an important tropical grain legume crop. Mungbean has the ability to fix atmospheric nitrogen and improve soil fertility (Mohammadzadeh et al. 2022). The seeds are rich in nutrients and bioactive compounds and are widely consumed as a food. Mungbean sprouts are a popular sprouting vegetable and are widely used as a fresh salad vegetable in Western countries on account of their high nutritional value and health benefits (Cheng et al. 2022).

High-quality reference genome sequences promote the study of genetic diversity in plants and are available for an increasing number of legume crops, such as soybean (*Glycine max* (L.) Merr.) (Shen et al. 2019), cowpea (*Vigna unguiculata* (L.) Walp.) (Lonardi et al. 2019), lima bean (*Phaseolus lunatus* L.) (Garcia et al. 2021), and mungbean (Liu et al. 2022). Owing to the substantial variation among individuals, a single reference genome cannot possibly contain all genetic information for a species (Golicz et al. 2016; Hurgobin et al. 2018). With advances in sequencing technology and genome-assembly methods, the pan-genome concept was introduced in a genomic analysis of multiple pathogenic isolates of a bacterial species. The pan-genome can represent complete genomic information for a species because it is constructed by sequencing dozens to hundreds of samples and enables progress in studying plant genomic evolution and identification of the critical genes associated with important agronomic traits (Li et al. 2022). In recent years, pan-genomic approaches have been successfully applied to many crop plants, such as rice (*Oryza sativa* L.) (Qin et al. 2021), maize (*Zea mays* L.) (Hufford et al. 2021), soybean (Liu et al. 2020; Torkamaneh et al. 2021; Gayali et al. 2016), and mungbean (Liu et al. 2022).

Given the rapid development of sequencing technology, numerous functional databases aggregating genomic, transcriptomic, proteomic, and germplasm data for leguminous crop species have been constructed and publicly released. CicerTransDB (Cicer Transcription Factor Database; http://www.cicertransdb.esy.es) can provide the complete transcription factors sequences in chickpea (*Cicer arietinum L.*) (Gayali et al. 2016). VigGS (*Vigna* Genome Server; http://viggs.dna.affrc.go.jp) contains gene structures and sequence polymorphisms visualized in GBrowse (Sakai et al. 2016). LegProt (http://bioinfo.noble.org/manuscript-support/legumedb)

enables protein identification by mass spectra (Lei et al. 2011). SoyBase is a genetic and genomic database compiled by the USDA-ARS (Brown et al. 2021; Grant et al. 2010). SoyTEdb provides resources and information on transposable elements in the soybean genome (Du et al. 2010). SoyNet is a database for co-functional networks (Kim et al. 2017). SoyProDB allows identification of innumerable proteins (Tavakolan et al. 2014). SoyProLow can be used to identify and characterize low-abundance proteins (Joshi et al. 2014). SoyKB provides information on genes/proteins, microRNAs/small RNAs, metabolites, single-nucleotide polymorphisms (SNPs) (Kang et al. 2014). These databases are valuable data sources for research on leguminous crop species. However, there is a lack of integration of the databases to facilitate comprehensive genomic analyses without the need to switch between databases.

In this study, we constructed the Legume Omics Resources Center (LORC) (https://njh.biobin.com.cn/cgi-bin/legume/Display.cgi?page=/home) to integrate genomic, transcriptomic, proteomic, and phenotypic data. LORC provides a wealth of multi-omics data and easy-to-use tools, and lays a solid foundation for analysis of the genetic diversity and breeding of legume crops.

2 Materials and Methods

2.1 Data Sources

To construct a comprehensive legume omics platform, we mined and integrated the following genomic, transcriptomic, and proteomic data. The Vrad JL7 de novo-assembled genome for Mungbean 'Jilv 7' (JL7) has a size of 475.19 Mb, predicts 40,125 protein-coding genes, and 42,986 transcripts are annotated. It is the highest-quality mungbean genome assembled to date (Liu et al. 2022). In total, genome assemblies from 217 mungbean accessions (including Vrad JL7) were aggregated to construct the mungbean pan-genome. Transcriptomic data for mungbean JL7 and 'VC1973A,' soybean, peanut (*Arachis hypogaea* L.), and wheat (*Triticum aestivum* L.) were accumulated from public databases (Liu et al. 2022; Emms and Kelly 2019).

2.2 Comparative Genomic Analysis

The sequences from the Vrad JL7 reference genome were compared with those of annotated genes from 12 eudicot plant species (*Arabidopsis thaliana* (L.) Heynh., *Arachis duranensis* Krapov. & W.C.Greg., *Cajanus cajan* (L.) Huth, *Cicer arietinum*, *Glycine max*, *Lotus japonicus* (Regel) K. Larsen, *Medicago truncatula* Gaertn., *Pisum sativum* L., *Phaseolus lunatus*, *Vigna angularis* (Willd.) Ohwi & H.Ohashi, *Phaseolus vulgaris* L., and *Vigna unguiculata*) to identify the characteristics of the mungbean genome. In the Vrad_JL7 genome, a total of 32,253 orthogroups/gene families (including 432 single-copy groups) were identified using OrthoFinder (version 2.37) (Danecek et al. 2011) and 35,059 coding genes were clustered into gene families.

2.3 SNPs and InDels

The genome resequencing data from 217 mungbean accessions were mapped to the Vrad_JL7 reference genome using BWA-MEM with the default parameters. Duplicate PCR products were removed by Picard (version 2.18.17). The Genome Analysis Toolkit (GATK; version 3.8.1) was used to detect genetic variants, including SNPs and short insertions/deletions (InDels; <15 bp) for each accession. The SNPs were filtered based on the following criteria: QD < 2.0, MQ < 40.0, FS > 60.0, SOR > 3.0, MQRankSum < −12.5, and ReadPosRankSum < −8.0. The SNPs/InDels with more than two alleles, missing rate > 10%, or minor allele frequency < 5% were removed by VCFtools (version 0.1.17). Based on the gene models of the Vrad_JL7 assembly, the genetic variants were annotated using SnpEff (version 5.0e) (Gao et al. 2019).

2.4 Pan-genome Construction and Annotation

The mungbean pan-genome was constructed as described previously (Zimin et al. 2013). For each of the 217 accessions, contigs were assembled using MaSuRCA (Gurevich et al. 2013). The assembled contigs were aligned to the Vrad JL7 genome using QUAST (Li and Godzik 2006). The contigs of length > 500 nt and identity less than the threshold of 90% were extracted as non-reference sequences. For the non-reference sequences, the sequences were aligned to the NT database and then the redundant sequences were removed by CD-HIT (version 4.8.1) (Li and Godzik 2006) The non-redundant contigs were aligned to the reference genome by BLAST. Finally, the Vrad_JL7, non-reference, and non-redundant contigs were merged as the reference pan-genome of mungbean.

The core and variable genes in the pan-genome were identified using a map-to-pan strategy. Mosdepth (version 0.3.1) was used to estimate coding-sequence coverage (Pedersen and Quinlan 2018). Genes with 2× coverage on at least 20% of the coding-sequence region were considered present, otherwise they were considered to be absent. The significance level for differences of gene numbers in different subgroups was tested using Wilcoxon's test ($P < 0.05$).

2.5 Genome-Wide Association Study (GWAS)

The SNPs and presence/absence variation (PAV) from 217 accessions of mungbean were filtered for GWAS. The GWAS was performed for 33 traits using GEMMA (version 0.98.1) (Zhou and Stephens 2012). Significant signals were identified based on the threshold $P < 0.05$ and by applying an adjusted Bonferroni test.

3 Results

3.1 Legume Omics Resources Center

The LORC online database integrates genomic, transcriptomic, proteomic, and phenotypic data and incorporates many analytical tools. LORC incorporates six functional tabs (Enter DB, BLAST, Tutorials, Browse, Analysis, and Contacts) at the top of the home page, a search bar on the upper left, and external links on the lower left (Fig. 1). These function labels provide convenient visual tools for users to easily view and compare gene annotations, genome structure, SNP markers, PAV information, and gene expression.

3.2 Mining and Browsing of LORC

In the search box, users can perform a search using a gene ID, EST/cDNA gene identification (gi) number, custom ID, or annotation to browse for gene information. When a gene ID is searched, LORC provides the results from six aspects: genome sequence, EST/cDNA information, GeneSeqer, intron information, transcripts, and summary. Users can switch between the tab pages to view each piece of information and tool usage (Case study 1).

Users click on the "Browse" button to enter the genome browser. In the genome browser, the reference genomes of Mungbean JL7, Mungbean Pub, adzuki bean (*Vigna angularis*), soybean, peanut, and wheat can be analyzed. Gene characteristic information, such as alternative splicing of ESTgi genes, and all expressed gene information are displayed by clicking to view the selected genome versions of the different species (Case study 1).

Case study 1 Mungbean JL7 is the most recent database assembled by Biobin for the Hebei Academy of Agriculture and Forestry Sciences. 'Jilv 7,' a high-yielding

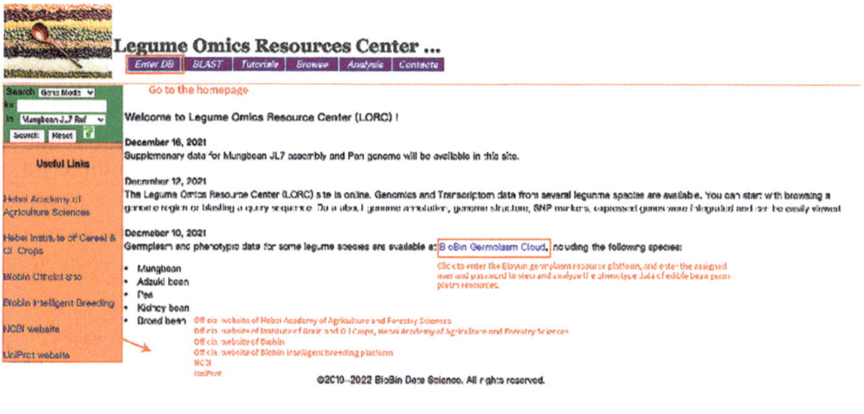

Fig. 1 Home page layout and data sources for the LORC website

and early-maturing mungbean cultivar widely cultivated in China, was sequenced using the PacBio Sequel II platform and Illumina paired-end (PE) read technology. Vrad_JL7, the de novo-assembled genome, has a size of 475.19 Mb and contains 40,125 predicted protein-coding genes and 42,986 annotated transcripts. It is the highest-quality mungbean genome assembly currently accessible. Here, we provide an example of using Mungbean JL7 for browsing in LORC. Users can input the gene ID, in this example g22646.t1, and LORC displays information for g22646.t1, including the genome, start and stop positions, and annotation (Fig. 2a). After clicking on the blue gene ID g22646.t1 in the left-hand column of the table, the coding region in the annotated protein, gene sequences aligned to other crop genomes, and corresponding RNA-seq and protein information are displayed (Fig. 2b). More information on g22646.t1 can be acquired by clicking on the different tabs, such as "Genome Sequence," "EST/cDNA," "Introns Info," and "Summary" (Fig. 2c, d). This example illustrates that the information for g22646.t1 is easy to locate and visualize, and thus that LORC can help researchers to quickly find and understand the features and functions of genes of interest

Users can also extract the results of alignment of a specified sequence for a species directly using BLAST. By clicking on the BLAST tab on the home page and selecting the item, e-value, database, and whether to filter, the comparison results will be automatically calculated and analyzed, and the analysis sequence provided.

3.3 Genome-Wide Association Analysis of Mungbean

To investigate the genetic diversity of mungbean, 217 mungbean accessions were resequenced several times. The average sequencing depth was 12× and ~ 2.23 M SNPs were identified in the 217 accessions. To identify the PAV of genes, the first mungbean pan-genome was constructed by aggregating the 217 mungbean accessions to generate an assembled pan-genome size of 763 Mb (including Vrad_JL7) and a total of 43,462 annotated genes.

A mungbean genome-wide association analysis system, for any SNP–trait association site (STA) and gene PAV of 33 agronomic traits in 217 mungbean accessions from different environments and years, can be viewed through a visual interface. The STAs were identified for all but one trait (trilobal leaf shape) in at least one environment via GWAS. A total of 2912 STAs were identified for each environment and 248 genes were located in the transcribed regions. STAs with multiple traits were clustered in hotspots of chromosomes 1, 4, and 7 in the mungbean genome, which may be associated with selection during mungbean domestication and improvement. In addition, 391 gene PAV–trait association (GPTA) events were identified from a gene PAV GWAS, which showed high consistency with STAs and may be complementary to the SNP method for identifying candidate genes (Song et al. 2020). All genome-wide associations can be queried and visually browsed in LORC (Fig. 3).

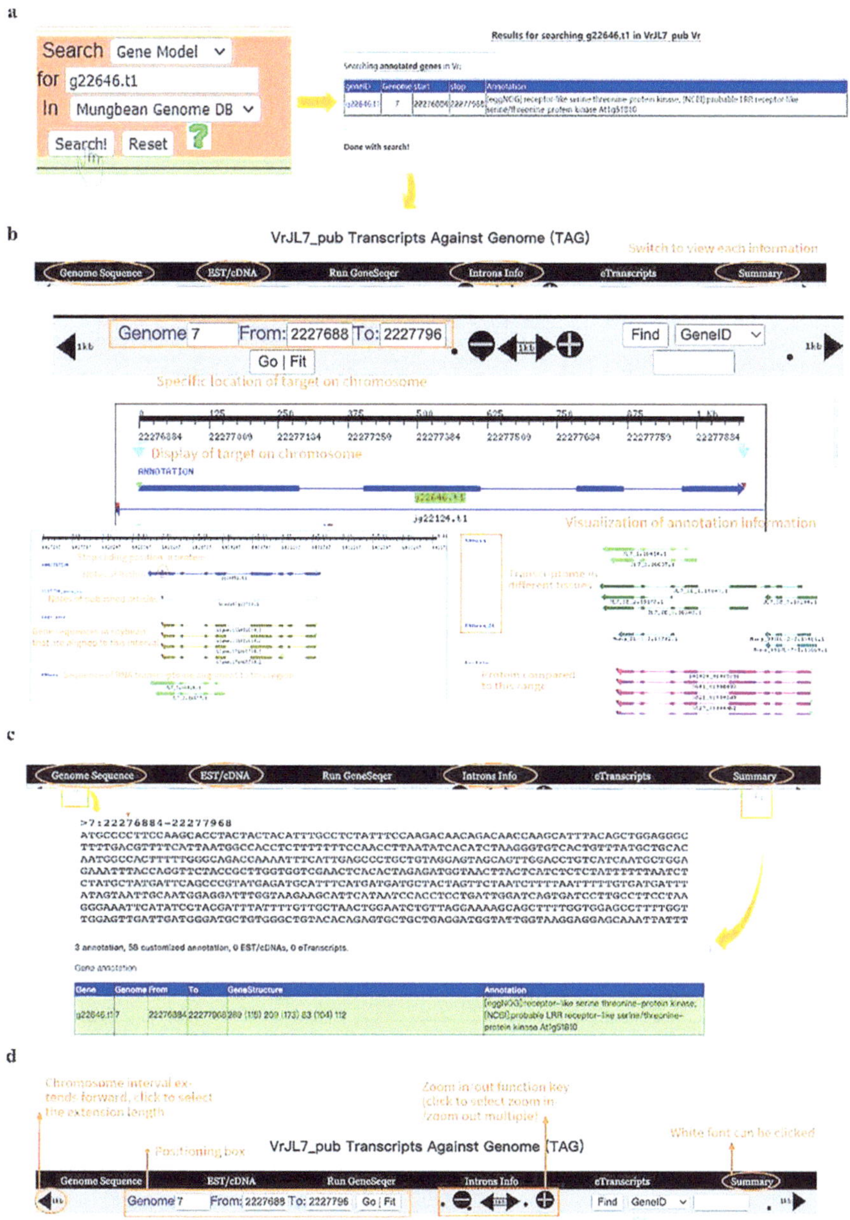

Fig. 2 Example of usage of the genome browser in LORC. (**a**) Information for the gene of interest; (**b**) visualization of the coding region in the annotated protein, gene sequences aligned to other crops, and RNA-sequencing and protein data; (**c**, **d**) display of "Genome Sequence," "EST/cDNA," "Introns Info," and "Summary" information for the gene of interest

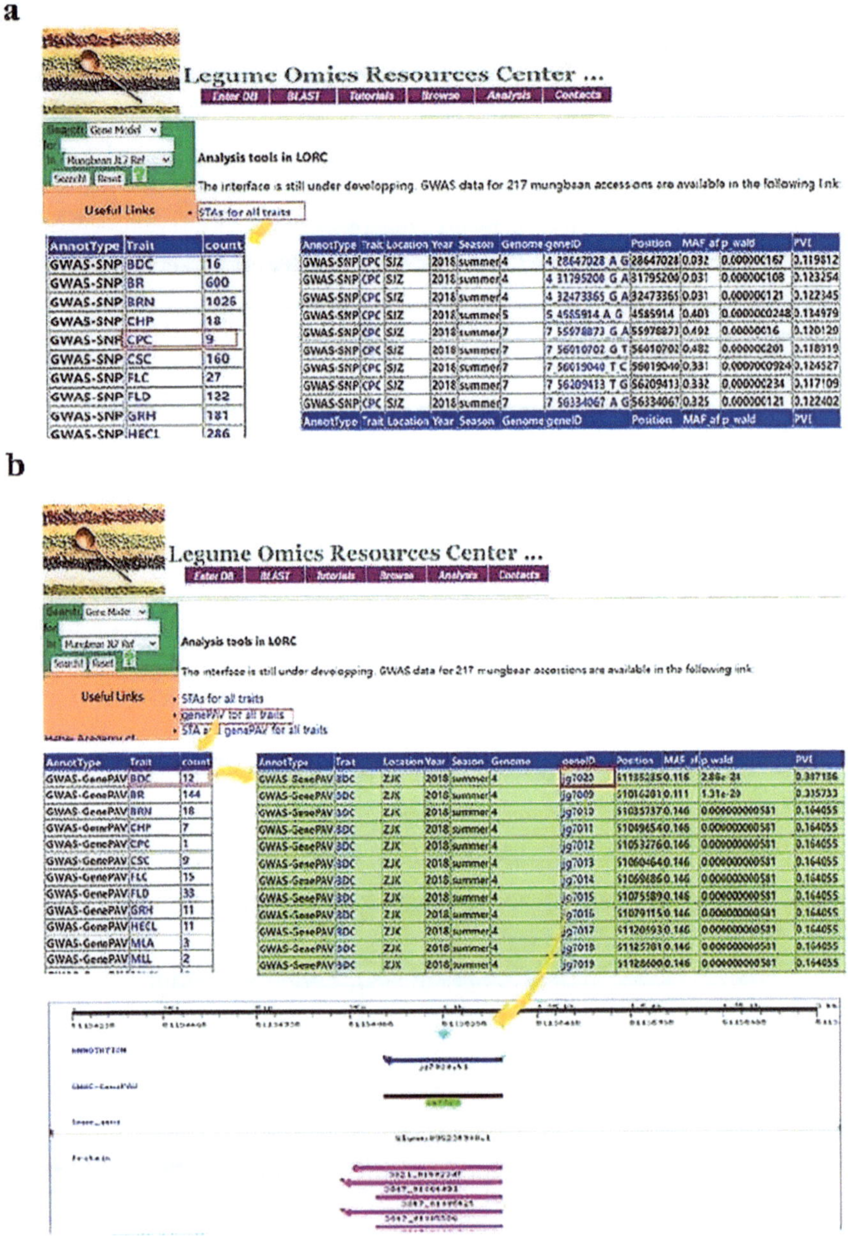

Fig. 3 Genome-wide association analysis of mungbean in LORC. (**a**) Site–trait associations (STAs) for all traits; (**b**) gene presence/absence variation (PAV) information for all traits

3.4 Bioinformatic Analysis Web Tools

To assist users in analyzing the data, LORC incorporates a number of common bioinformatic web tools, including BLAST, genome resequencing analysis, and transcriptome analysis. Thus, users without a bioinformatic background can perform an analysis of the data and are not required to switch between different databases or modules (Case study 2).

Case study 2 LORC contains a wealth of information, such as multi-omics data, homologous alignments, and the BLAST tool. LORC integrates genomic and transcriptomic data for several common legume crops and supports multi-entry and multiple ID types in a search (Fig. 4a). LORC provides detailed information on IDs in a table, such as gene annotations, genome structure, and expressed genes, and queries five alternative splicing conditions of transcripts from legume crops (except for Mungbean JL7 and adzuki bean) (Fig. 4b, c). It also provides multiple backgrounds (including Mungbean VrJL7 and soybean genome assemblies, and the NCBI nt and nr databases) in BLAST and visualizes the results of the BLAST search (Fig. 4e)

4 Discussion

In this study, we mined and integrated genomic, transcriptomic, genetic variation, phenotypic, and epigenomic data for common legume crops. Compared with existing legume databases, including CicerTransDB, VigGS, LegProt, SoyBase, SoyTEdb, SoyNet, SoyProDB, SoyProLow, and SoyKB, LORC has a number of attractive advantages (Table 1). LORC is the first database to provide genome-wide association analysis of mungbean, which is crucial to understand the genomic diversity. LORC provides the online multi-omics analysis platform for selected legume crops. LORC provides multiple search and visualization functions, including genomic sequences, genome structure, gene annotations, single-nucleotide polymorphism markers, presence/absence variation information, and gene expression data. LORC integrates genomic, transcriptomic, proteomic, and phenotypic data. It provides some easy-to-use tools, laying a solid foundation for analysis of the genetic diversity and breeding of legume crops.

In the future, resequencing data and genome assemblies for an increasing number of species will become available to enable rapid analysis of the genetic variation and genes associated with important traits. With further technological developments, a greater number of datasets will be available. Hence, it will be critical to integrate multi-omics data. We will constantly update and integrate omics data from additional accessions to improve LORC.

Fig. 4 Bioinformatic web tools implemented in LORC. (**a**) LORC supports searches involving multiple entries and multiple ID types. (**b, c**) LORC provides detailed information on the relevant gene IDs in an output table. (**d, e**) LORC provides multiple backgrounds for BLAST searches. The red arrows point to the next

Table 1 Summary of features distinguishing LORC from existing legume species genetic databases

Database	Major function	Species
CicerTransDB	Transcription factor database, manual classification of TF families, gene information, gene ontology, domain and motif architecture.	*Cicer arietinum*
'VigGS'	Gene structures, sequence polymorphisms, genome sequencing of 13 wild *Vigna* species	*Vigna angularis*
LegProt	Protein identification	*Glycine max,* *Lotus japonicus Medicago sativa* *Medicago truncatula Lupinu salbus Phaseolus vulgaris* *Pisum sativum*
SoyBase	Genetics and genomics date	*Glycine max*
SoyTEdb	Transposable elements, provides resources and information related to transposable elements in the soybean genome	*Glycine max*
SoyNet	Co-functional networks	*Glycine max*
SoyProDB	Seed proteins, identify several proteins	*Glycine max*
SoyProLow	Low-abundance proteins, identify and characterize low abundance proteins	*Glycine max*
SoyKB	Translational genomics and molecular breeding	*Glycine max*
LORC	Multi-omics data resource and analysis tools	*Vigna radiata* *Glycine max* *Vigna angularis* *Pisum sativum Phaseolus vulgaris Vicia faba* *Arachis hypogaea Triticum aestivum*

5 Conclusion

LORC contains a wealth of genetic information, such as multi-omics data, homologous alignments, and the BLAST sequence analysis tool. LORC provides important resources and convenient visualization tools for researchers to retrieve required information online, which will accelerate research progress and promote the rapid genetic improvement of legume crops. Further application and extension of LORC are expected to contribute to improved mungbean production, breeding, and expansion of the mungbean market.

Acknowledgments The authors appreciate researchers from 20 multi-ecological identification sites for providing data. We thank Robert McKenzie, PhD, from Liwen Bianji (Edanz) (www.liwenbianji.cn) for editing a draft of this manuscript.

Data Availability Statement The datasets used and/or analyzed during the current study are available from the authors on reasonable request.

Funding This research was funded by the National Key R&D Program of China (2019YFD1000704-3), the China Agriculture Research System (CARS-08-G3), the Hebei Innovation Team of Modern Seed Industry (21326305D), and the Hebei Agriculture Research System (HBCT2023050205).

References

Brown AV, Conners SI, Huang W, Wilkey AP, Grant D, Weeks NT, Cannon SB, Graham MA, Nelson RT (2021) A new decade and new data at SoyBase, the USDA-ARS soybean genetics and genomics database. Nucleic Acids Res 49(D1):D1496–D1501. https://doi.org/10.1093/nar/gkaa1107

Cheng Y, Xiang N, Cheng X, Chen H, Guo X (2022) Effect of photoperiod on polyphenol biosynthesis and cellular antioxidant capacity in mung bean (*Vigna* radiata) sprouts. Food Res Int 159:111626. https://doi.org/10.1016/j.foodres.2022.111626

Danecek P, Auton A, Abecasis G, Albers CA, Banks E, DePristo MA, Handsaker RE, Lunter G, Marth GT, Sherry ST, McVean G (2011) The variant call format and VCFtools. Bioinformatics 27(15):2156–2158. https://doi.org/10.1093/bioinformatics/btr330

Du J, Grant D, Tian Z, Nelson RT, Zhu L, Shoemaker RC, Ma J (2010) SoyTEdb: a comprehensive database of transposable elements in the soybean genome. BMC Genomics 11(1):113. https://doi.org/10.1186/1471-2164-11-113

Emms DM, Kelly S (2019) OrthoFinder: phylogenetic orthology inference for comparative genomics. Genome Biol 20(1):238. https://doi.org/10.1186/s13059-019-1832-y

Gao L, Gonda I, Sun H, Ma Q, Bao K, Tieman DM, Burzynski-Chang EA, Fish TL, Stromberg KA, Sacks GL, Thannhauser TW (2019) The tomato pan-genome uncovers new genes and a rare allele regulating fruit flavor. Nat Genet 51(6):1044–1051. https://doi.org/10.1038/s41588-019-0410-2

Garcia T, Duitama J, Zullo SS, Gil J, Ariani A, Dohle S, Palkovic A, Skeen P, Bermudez-Santana CI, Debouck DG, Martínez-Castillo J (2021) Comprehensive genomic resources related to domestication and crop improvement traits in Lima bean. Nat Commun 12(1):702. https://doi.org/10.1038/s41467-021-20921-1

Gayali S, Acharya S, Lande NV, Pandey A, Chakraborty S, Chakraborty N (2016) CicerTransDB 1.0: a resource for expression and functional study of chickpea transcription factors. BMC Plant Biol 16(1):169. https://doi.org/10.1186/s12870-016-0860-y

Golicz AA, Batley J, Edwards D (2016) Towards plant pangenomics. Plant Biotechnol J 14(4):1099–1105. https://doi.org/10.1111/pbi.12499

Grant D, Nelson RT, Cannon SB, Shoemaker RC (2010) SoyBase, the USDA-ARS soybean genetics and genomics database. Nucleic Acids Res 38(suppl_1):D843–D846. https://doi.org/10.1093/nar/gkp798

Gurevich A, Saveliev V, Vyahhi N, Tesler G (2013) QUAST: quality assessment tool for genome assemblies. Bioinformatics 29(8):1072–1075. https://doi.org/10.1093/bioinformatics/btt086

Hufford MB, Seetharam AS, Woodhouse MR, Chougule KM, Ou S, Liu J, Ricci WA, Guo T, Olson A, Qiu Y, Della Coletta R (2021) De novo assembly, annotation, and comparative analysis of 26 diverse maize genomes. Science 373(6555):655–662. https://doi.org/10.1126/science.abg5289

Hurgobin B, Golicz AA, Bayer PE, Chan CK, Tirnaz S, Dolatabadian A, Schiessl SV, Samans B, Montenegro JD, Parkin IA, Pires JC (2018) Homoeologous exchange is a major cause of gene presence/absence variation in the amphidiploid Brassica napus. Plant Biotechnol J 16(7):1265–1274. https://doi.org/10.1111/pbi.12867

Joshi T, Fitzpatrick MR, Chen S, Liu Y, Zhang H, Endacott RZ, Gaudiello EC, Stacey G, Nguyen HT, Xu D (2014) Soybean knowledge base (SoyKB): a web resource for integration of soybean translational genomics and molecular breeding. Nucleic Acids Res 42(D1):D1245–D1252. https://doi.org/10.1093/nar/gkt905

Kang YJ, Kim SK, Kim MY, Lestari P, Kim KH, Ha BK, Jun TH, Hwang WJ, Lee T, Lee J, Shim S (2014) Genome sequence of Mungbean and insights into evolution within *Vigna* species. Nat Commun 5(1):5443. https://doi.org/10.1038/ncomms6443

Kim E, Hwang S, Lee I (2017) SoyNet: a database of co-functional networks for soybean Glycine max. Nucleic Acids Res 45(D1):D1082–D1089. https://doi.org/10.1093/nar/gkw704

Lei Z, Dai X, Watson BS, Zhao PX, Sumner LW (2011) A legume specific protein database (LegProt) improves the number of identified peptides, confidence scores and overall protein identification success rates for legume proteomics. Phytochemistry 72(10):1020–1027. https://doi.org/10.1016/j.phytochem.2011.01.026

Li W, Godzik A (2006) Cd-hit: a fast program for clustering and comparing large sets of protein or nucleotide sequences. Bioinformatics 22(13):1658–1659. https://doi.org/10.1007/978-1-4614-6418-1_221-2

Li W, Liu J, Zhang H, Liu Z, Wang Y, Xing L, He Q, Du H (2022) Plant pan-genomics: recent advances, new challenges, and roads ahead. J Genet Genomics 49(9):833–846. https://doi.org/10.1016/j.jgg.2022.06.004

Liu Y, Du H, Li P, Shen Y, Peng H, Liu S, Zhou GA, Zhang H, Liu Z, Shi M, Huang X (2020) Pan-genome of wild and cultivated soybeans. Cell 182(1):162–176. https://doi.org/10.1016/j.cell.2020.05.023

Liu C, Wang Y, Peng J, Fan B, Xu D, Wu J, Cao Z, Gao Y, Wang X, Li S, Su Q (2022) High-quality genome assembly and pan-genome studies facilitate genetic discovery in mung bean and its improvement. Plant Commun 3(6). https://doi.org/10.1016/j.xplc.2022.100352

Lonardi S, Muñoz-Amatriaín M, Liang Q, Shu S, Wanamaker SI, Lo S, Tanskanen J, Schulman AH, Zhu T, Luo MC, Alhakami H (2019) The genome of cowpea (*Vigna* unguiculata [L.] Walp.). Plant J 98(5):767–782. https://doi.org/10.1111/tpj.14349

Mohammadzadeh V, Rezaei-Chiyaneh E, Mahdavikia H, Rahimi A, Gheshlaghi M, Battaglia ML, Harrison MT (2022) Effect of intercropping and bio-fertilizer application on the nutrient uptake and productivity of mung bean and marjoram. Land 11(10):1825. https://doi.org/10.3390/land11101825

Pedersen BS, Quinlan AR (2018) Mosdepth: quick coverage calculation for genomes and exomes. Bioinformatics 34(5):867–868. https://doi.org/10.1093/bioinformatics/btx699

Qin P, Lu H, Du H, Wang H, Chen W, Chen Z, He Q, Ou S, Zhang H, Li X, Li X (2021) Pan-genome analysis of 33 genetically diverse rice accessions reveals hidden genomic variations. Cell 184(13):3542–3558. https://doi.org/10.1016/j.cell.2021.04.046

Sakai H, Naito K, Takahashi Y, Sato T, Yamamoto T, Muto I, Itoh T, Tomooka N (2016) The *Vigna* Genome Server,'Vig GS': A genomic knowledge base of the genus *Vigna* based on high-quality, annotated genome sequence of the Azuki Bean, *Vigna angularis* (Willd.) Ohwi & Ohashi. Plant Cell Physiol 57(1):e2. https://doi.org/10.1093/pcp/pcv189

Shen Y, Du H, Liu Y, Ni L, Wang Z, Liang C, Tian Z (2019) Update soybean Zhonghuang 13 genome to a golden reference. Sci China Life Sci 62(9):1257–1260. https://doi.org/10.1007/s11427-019-9822-2

Song JM, Guan Z, Hu J, Guo C, Yang Z, Wang S, Liu D, Wang B, Lu S, Zhou R, Xie WZ (2020) Eight high-quality genomes reveal pan-genome architecture and ecotype differentiation of Brassica napus. Nat Plants 6(1):34–45. https://doi.org/10.1038/s41477-019-0577-7

Tavakolan M, Alkharouf NW, Matthews BF, Natarajan SS (2014) SoyProLow: a protein database enriched in low abundant soybean proteins. Bioinformation 10(9):599. https://doi.org/10.6026/97320630010599

Torkamaneh D, Lemay MA, Belzile F (2021) The pan-genome of the cultivated soybean (PanSoy) reveals an extraordinarily conserved gene content. Plant Biotechnol J 19(9):1852–1862. https://doi.org/10.1111/pbi.13600

Zhou X, Stephens M (2012) Genome-wide efficient mixed-model analysis for association studies. Nat Genet 44:821–824. https://doi.org/10.1038/ng.2310

Zimin AV, Marçais G, Puiu D, Roberts M, Salzberg SL, Yorke JA (2013) The MaSuRCA genome assembler. Bioinformatics 29(21):2669–2677. https://doi.org/10.1093/bioinformatics/btt476

MIX
Papier aus verantwortungsvollen Quellen
Paper from responsible sources
FSC® C105338

If you have any concerns about our products,
you can contact us on
ProductSafety@springernature.com

In case Publisher is established outside the EU,
the EU authorized representative is:
Springer Nature Customer Service Center GmbH
Europaplatz 3, 69115 Heidelberg, Germany

Printed by Libri Plureos GmbH
in Hamburg, Germany